INVESTIGATIONS IN NUMBER, DATA, AND SPACE®

Exploring Geometry

Making Shapes and Building Blocks

Kindergarten
Also appropriate for Grade 1

Karen Economopoulos
Megan Murray
Kim O'Neil
Doug Clements
Julie Sarama
Susan Jo Russell

Developed at TERC, Cambridge, Massachusetts

Dale Seymour Publications®
White Plains, New York

The *Investigations* curriculum was developed at TERC (formerly Technical Education Research Centers) in collaboration with Kent State University and the State University of New York at Buffalo. The work was supported in part by National Science Foundation Grant No. ESI-9050210. TERC is a nonprofit company working to improve mathematics and science education. TERC is located at 2067 Massachusetts Avenue, Cambridge, MA 02140.

This project was supported, in part, by the
National Science Foundation
Opinions expressed are those of the authors and not necessarily those of the Foundation

Managing Editor: Catherine Anderson
Series Editor: Beverly Cory
ESL Consultant: Nancy Sokol Green
Production/Manufacturing Director: Janet Yearian
Production/Manufacturing Manager: Karen Edmonds
Production/Manufacturing Coordinator: Roxanne Knoll
Design Manager: Jeff Kelly
Design: Don Taka
Illustrations: DJ Simison, Carl Yoshihara
Composition: Joe Conte
Cover: Bay Graphics

Shapes™ program: copyright © 1997, D. H. Clements
Logo core is © LCSI, 1993
Shapes activities: copyright © 1998, Dale Seymour Publications®
Apple and Macintosh are registered trademarks of Apple Computer, Inc.

This book is published by Dale Seymour Publications®, an imprint of Addison Wesley Longman, Inc.

> Dale Seymour Publications
> 10 Bank Street
> White Plains, NY 10602
> Customer Service: 1-800-872-1100

Copyright © 1998 by Dale Seymour Publications®. All rights reserved. Printed in the United States of America.

Limited reproduction permission: The publisher grants permission to individual teachers who have purchased this book to reproduce the blackline masters as needed for use with their own students. Reproduction for an entire school or school district or for commercial use is prohibited.

Printed on Recycled Paper

Order number DS47106
ISBN 1-57232-929-7

1 2 3 4 5 6 7 8 9 10-ML-02 01 00 99 98

TERC

INVESTIGATIONS IN NUMBER, DATA, AND SPACE®

Principal Investigator   Susan Jo Russell
Co-Principal Investigator   Cornelia Tierney
Director of Research and Evaluation   Jan Mokros
Director of K–2 Curriculum   Karen Economopoulos

Curriculum Development
Karen Economopoulos
Rebeka Eston
Marlene Kliman
Christopher Mainhart
Jan Mokros
Megan Murray
Kim O'Neil
Susan Jo Russell
Tracey Wright

Evaluation and Assessment
Mary Berle-Carman
Jan Mokros
Andee Rubin

Teacher Support
Irene Baker
Megan Murray
Kim O'Neil
Judy Storeygard
Tracey Wright

Technology Development
Michael T. Battista
Douglas H. Clements
Julie Sarama

Video Production
David A. Smith
Judy Storeygard

Administration and Production
Irene Baker
Amy Catlin

Cooperating Classrooms for This Unit
Jeanne Wall
Arlington Public Schools
Arlington, MA

Audrey Barzey
Patricia Kelliher
Ellen Tait
Boston Public Schools
Boston, MA

Meg Bruton
Fayerweather Street School
Cambridge, MA

Rebeka Eston
Lincoln Public Schools
Lincoln, MA

Lila Austin
The Atrium School
Watertown, MA

Christopher Mainhart
Westwood Public Schools
Westwood, MA

Consultants and Advisors
Deborah Lowenberg Ball
Michael T. Battista
Marilyn Burns
Douglas H. Clements
Ann Grady

CONTENTS

About the *Investigations* Curriculum	I-1
How to Use This Book	I-2
Technology in the Curriculum	I-6
About Assessment	I-8

Making Shapes and Building Blocks

Unit Overview	I-10
Materials List	I-16
About the Mathematics in This Unit	I-17
About the Assessment in This Unit	I-18
Preview for the Linguistically Diverse Classroom	I-20

Investigation 1: 2-D Shapes Around Us	2
Investigation 2: Exploring Shapes with the Computer	26
Investigation 3: Looking at 3-D Shapes	38
Investigation 4: Making Shapes and Building Blocks	60
Investigation 5: 2-D Faces on 3-D Blocks	82

Choice Time Activities

Book of Shapes	12
Pattern Block Pictures	14
Shape Mural	17
Free Explore with *Shapes* on the Computer	31
Pattern Block Puzzles	34
Shape Hunt	46
Exploring Geoblocks	48
The Shape of Things on the Computer	50
Clay Shapes	68
Fill the Hexagons	70
Build a Block	73
Quick Images on the Computer	75
Matching Faces	88
Geoblock Match-Up	90
Planning Pictures on the Computer	92

General Teacher Notes	96
About Classroom Routines	103
Tips for the Linguistically Diverse Classroom	114
Vocabulary Support for Second-Language Learners	116
Shapes Teacher Tutorial	117
Blackline Masters: Family Letter, Student Sheets, Teaching Resources	155

TEACHER NOTES

Learning to See Shapes in the Environment	21
How Young Children Learn About Shapes	22
About Pattern Blocks	24
Managing the Computer Activities	36
Why Use Pattern Block Shapes on the Computer?	37
About Geoblocks	55
Making Shapes	79
About Choice Time	96
Materials as Tools for Learning	99
Encouraging Students to Think, Reason, and Share Ideas	100
Games: The Importance of Playing More Than Once	101

WHERE TO START

The first-time user of *Making Shapes and Building Blocks* should read the following:

■ About the Mathematics in This Unit	I-17
■ About the Assessment in This Unit	I-18
■ Teacher Note: How Young Children Learn About Shapes	22
■ Teacher Note: About Geoblocks	55
■ Teacher Note: Making Shapes	79

If you plan to use computers as you teach this unit, read the following:

■ Teacher Note: Managing the Computer Activities	36
■ Teacher Note: Why Use Pattern Block Shapes on the Computer?	37
■ *Shapes* Teacher Tutorial	117

When you next teach this same unit, you can begin to read more of the background. Each time you present the unit, you will learn more about how your students understand the mathematical ideas.

ABOUT THE *INVESTIGATIONS* CURRICULUM

Investigations in Number, Data, and Space® is a K–5 mathematics curriculum with four major goals:

- to offer students meaningful mathematical problems
- to emphasize depth in mathematical thinking rather than superficial exposure to a series of fragmented topics
- to communicate mathematics content and pedagogy to teachers
- to substantially expand the pool of mathematically literate students

The *Investigations* curriculum embodies a new approach based on years of research about how children learn mathematics. Each grade level consists of a set of separate units, each offering 2–8 weeks of work. These units of study are presented through investigations that involve students in the exploration of major mathematical ideas.

Approaching the mathematics content through investigations helps students develop flexibility and confidence in approaching problems, fluency in using mathematical skills and tools to solve problems, and proficiency in evaluating their solutions. Students also build a repertoire of ways to communicate about their mathematical thinking, while their enjoyment and appreciation of mathematics grows.

The investigations are carefully designed to invite all students into mathematics—girls and boys, members of diverse cultural, ethnic, and language groups, and students with different strengths and interests. Problem contexts often call on students to share experiences from their family, culture, or community. The curriculum eliminates barriers—such as work in isolation from peers, or emphasis on speed and memorization—that exclude some students from participating successfully in mathematics. The following aspects of the curriculum ensure that all students are included in significant mathematics learning:

- Students spend time exploring problems in depth.
- They find more than one solution to many of the problems they work on.
- They invent their own strategies and approaches, rather than rely on memorized procedures.
- They choose from a variety of concrete materials and appropriate technology, including calculators, as a natural part of their everyday mathematical work.
- They express their mathematical thinking through drawing, writing, and talking.
- They work in a variety of groupings—as a whole class, individually, in pairs, and in small groups.
- They move around the classroom as they explore the mathematics in their environment and talk with their peers.

While reading and other language activities are typically given a great deal of time and emphasis in elementary classrooms, mathematics often does not get the time it needs. If students are to experience mathematics in depth, they must have enough time to become engaged in real mathematical problems. We believe that a minimum of 5 hours of mathematics classroom time a week—about an hour a day—is critical at the elementary level. The scope and pacing of the *Investigations* curriculum are based on that belief.

We explain more about the pedagogy and principles that underlie these investigations in Teacher Notes throughout the units. For correlations of the curriculum to the NCTM Standards and further help in using this research-based program for teaching mathematics, see the following books, available from Dale Seymour Publications:

- *Implementing the* Investigations in Number, Data, and Space® *Curriculum*
- *Beyond Arithmetic: Changing Mathematics in the Elementary Classroom* by Jan Mokros, Susan Jo Russell, and Karen Economopoulos

HOW TO USE THIS BOOK

This book is one of the curriculum units for *Investigations in Number, Data, and Space*. In addition to providing part of a complete mathematics curriculum for your students, this unit offers information to support your own professional development. You, the teacher, are the person who will make this curriculum come alive in the classroom; the book for each unit is your main support system.

Although the curriculum does not include student instructional texts, reproducible sheets for student work are provided with the units and, in some cases, are also available as Student Activity Booklets. In these investigations, students work actively with objects and experiences in their own environment, including manipulative materials and technology, rather than with a workbook.

Ultimately, every teacher will use these investigations in ways that make sense for his or her particular style, the particular group of students, and the constraints and supports of a particular school environment. Each unit offers information and guidance drawn from our collaborations with many teachers and students over many years. Our goal is to help you, a professional educator, give all your students access to mathematical power.

Investigation Format

The opening two pages of each investigation help you get ready for the work that follows.

- **Focus Time** This gives a synopsis of the activities used to introduce the important mathematical ideas for the investigation.
- **Choice Time** This lists the activities, new and recurring, that support the Focus Time work.
- **Mathematical Emphasis** This highlights the most important ideas and processes students will encounter in this investigation.
- **Teacher Support** This indicates the Teacher Notes and Dialogue Boxes included to help you understand what's going on mathematically in your classroom.
- **What to Plan Ahead of Time** These lists alert you to materials to gather, sheets to duplicate, and other things you need to do before starting the investigation. Full details of materials and preparation are included with each activity.

INVESTIGATION 1

2-D Shapes Around Us

Focus Time

Looking at 2-D Shapes (p. 4)
Students begin to think about two-dimensional shapes as they look for and talk about the shapes around their classroom. Together they look at a book about the shape of things in the world and then design individual pages for a class Book of Shapes.

Choice Time

Book of Shapes (p. 12)
Students continue to work on their pages for the class Book of Shapes.

Pattern Block Pictures (p. 14)
After exploring pattern blocks, students make a permanent representation of their pattern block designs or pictures.

Shape Mural (p. 17)
Students work together to design and make a shape mural, using cutout paper shapes to represent objects and people.

A Note on Using Computers

Shapes, a software program that enables students to work with 2-D shapes on the computer, is introduced in Investigation 2. The **Teacher Note**, Managing the Computer Activities (p. 36), discusses options for integrating computer work with this unit. If you will be using computers, you may want to merge Investigations 1 and 2, introducing the computer to small groups while the rest of the class works on Choice Time. If computers are not available, plan to skip Investigation 2.

Mathematical Emphasis

- Recognizing shapes in the environment
- Observing and describing two-dimensional (2-D) shapes
- Developing vocabulary to describe 2-D shapes
- Becoming familiar with the names of 2-D shapes
- Relating 2-D shapes to real-world objects

Teacher Support

Teacher Notes
Learning to See Shapes in the Environment (p. 21)
How Young Children Learn About Shapes (p. 22)
About Pattern Blocks (p. 24)

Dialogue Box
Ideas for a Book of Shapes (p. 25)

"A square can be a present."

INVESTIGATION 1

What to Plan Ahead of Time

Focus Time Materials

Looking at 2-D Shapes
- *The Shape of Things* by Dayle Ann Dodds (Candlewick Press, 1994) or a similar book about shapes in the environment (see p. 4 for alternative titles)
- Shape Cutouts A–F (pp. 157–159): about 10 of each, copied on paper of different colors
- Large paper (11 by 17 inches): 1 sheet per student, plus extras
- Crayons, markers, and/or colored pencils
- Art supplies, such as stencils, rubber stamps, and stickers
- Glue sticks or paste

Choice Time Materials

Book of Shapes
- Shape Cutouts in a variety of colors, large paper, coloring materials, art supplies, and glue sticks from Focus Time
- A stapler or hole-punch and string, for binding the class book (optional)

Pattern Block Pictures
- Pattern blocks: 1 bucket per 4–6 students
- Small cups or containers to use as scoops (optional)
- Paper pattern blocks: 1 manufactured set, or prepare from masters on pp. 160–165
- Unlined paper: 1 sheet per student
- Glue sticks

Shape Mural
- *The Shape of Things* by Dayle Ann Dodds (or similar book of shapes)
- Butcher paper, oak tag, or other large paper for a mural
- Paper pattern blocks: 1 manufactured set, or prepare from masters on pp. 160–165
- Shapes Cutouts remaining from work on the Book of Shapes (make extras as needed)
- Coloring materials and art supplies
- Glue sticks or paste
- Chart paper (optional)

Family Connection
- Family letter (p. 156) or *Investigations at Home*: 1 per family

Always read through an entire investigation before you begin, in order to understand the overall flow and sequence of the activities.

Focus Time In this whole-group meeting, you introduce one or more activities that embody the important mathematical ideas underlying the investigation. The group then may break up into individuals or pairs for further work on the same activity. Many Focus Time activities culminate with a brief sharing time or discussion as a way of acknowledging students' work and highlighting the mathematical ideas. Focus Time varies in length. Sometimes it is short and can be completed in a single group meeting or a single work period; other times it may stretch over two or three sessions.

Choice Time Each Focus Time is followed by Choice Time, which offers a series of supporting activities to be done simultaneously by individuals, pairs, or small groups. You introduce these related tasks over a period of several days. During Choice Time, students work independently, at their own pace, choosing the activities they prefer and often returning many times to their favorites. Many kindergarten classrooms have an activity time built into their daily schedule, and Choice Time activities can easily be incorporated.

Together, the Focus Time and Choice Time activities offer a balanced kindergarten curriculum.

Classroom Routines The kindergarten day is filled with opportunities to work with mathematics. Routines such as taking attendance, asking about snack preferences, and discussing the calendar offer regular, ongoing practice in counting, collecting and organizing data, and understanding time.

Four specific routines—Attendance, Counting Jar, Calendar, and Today's Question—are formally introduced in the unit *Mathematical Thinking in Kindergarten*. Another routine, Patterns on the Pocket Chart, is introduced in the unit *Pattern Trains and Hopscotch Paths*. Descriptions of these routines can also be found in an appendix for each unit, and reminders of their ongoing use appear in the Unit Overview charts.

The Linguistically Diverse Classroom Each unit includes an appendix with Tips for the Linguistically Diverse Classroom to help teachers support students at varying levels of English proficiency. While more specific tips appear within the units at grades 1–5, often in relation to written work, general tips on oral discussions and observing the students are more appropriate for kindergarten.

Also included are suggestions for vocabulary work to help ensure that students' linguistic difficulties do not interfere with their comprehension of math concepts. The Preview for the Linguistically Diverse Classroom lists key words in the unit that are generally known to English-speaking kindergartners. Activities to help familiarize other students with these words are found in the appendix, Vocabulary Support for Second-Language Learners. In addition, ideas for making connections to students' languages and cultures, included on the Preview page, help the class explore the unit's concepts from a multicultural perspective.

Materials

A complete list of the materials needed for teaching this unit follows the Unit Overview. These materials are available in *Investigations* kits or can be purchased from school supply dealers.

Classroom Materials In an active kindergarten mathematics classroom, certain basic materials should be available at all times, including interlocking cubes, a variety of things to count with, and writing and drawing materials. Some activities in this curriculum require scissors and glue sticks or tape; dot stickers and large paper are also useful. So that students can independently get what they need at any time, they should know where the materials are kept, how they are stored, and how they are to be returned to the storage area.

Children's Literature Each unit offers a list of children's literature that can be used to support the mathematical ideas in the unit. Sometimes an activity incorporates a specific children's book, with suggestions for substitutions where practical. While such activities can be adapted and taught without the book, the literature offers a rich introduction and should be used whenever possible. If you can get the titles in Big Book format, these are ideal for kindergarten.

Blackline Masters Student recording sheets and other teaching tools for both class and homework are provided as reproducible blackline masters at the end of each unit. When student sheets are designated for homework at kindergarten, they usually repeat an activity from class, such as playing a game, as a way of involving and informing family members. Occasionally a homework sheet may ask students to collect data or materials for a class project or in preparation for upcoming activities.

Student Activity Booklets For the two kindergarten number units, the blackline masters are also available as Student Activity Booklets, designed to free you from extensive copying. The other kindergarten units require minimal copying.

Family Letter A letter that you can send home to students' families is included with the blackline masters for each unit. Families need to be informed about the mathematics work in your classroom; they should be encouraged to participate in and support their children's work. A reminder to send home the letter for each unit appears in one of the early investigations. These letters are also available separately in Spanish, Vietnamese, Cantonese, Hmong, and Cambodian.

***Investigations* at Home** To further involve families in the kindergarten program, you can offer them the *Investigations* at Home booklet, which describes the kindergarten units, explains the mathematics work children do in kindergarten, and offers activities families can do with their children at home.

Adapting *Investigations* to Your Classroom

Kindergarten programs vary greatly in the amount of time each day that students attend. We recommend that kindergarten teachers devote from 30 to 45 minutes daily to work in mathematics, but we recognize that this can be challenging in a half-day program. The kindergarten level of *Investigations* is intentionally flexible so that teachers can adapt the curriculum to their particular setup.

Kindergartens participating in the *Investigations* field test included full-day programs, half-day programs of approximately 3 hours, and half-day programs that add one or two full days to the kindergarten week at some point in the school year. Despite the wide range of program structures, classrooms generally fell into one of two groups: those that offered a separate math time daily (Math Workshop or Math Time), and those that included one or two mathematics activities during a general Activity Time or Station Time.

Math Workshop Teachers using a Math Workshop approach set aside 30 to 45 minutes each day for doing mathematics. In addition, they usually also have a more general activity time in their daily schedule. On some days, Math Workshop might be devoted to the Focus Time activities, with the whole class gathered together. On other days, students might work in small groups and choose from three or four Choice Time activities.

Math as Part of Activity Time Teachers with less time in their day may offer students one or two math activities, along with activities from other areas of the curriculum, during their Activity Time or Station Time. For example, on a particular day, students might be able to choose among a science activity, block building, an art project, dramatic play, books, puzzles, and a math activity. New activities are introduced during a whole-class meeting. With the *Investigations* curriculum, teachers who use this approach have found that it is important to designate at least one longer block of time (30 to 45 minutes) each week for mathematics. During this time, students engage in Focus Time activities and have a chance to share their work and discuss mathematical ideas. The suggested Choice Time activities are then presented as part of the general activity time. Following this model, work on a curriculum unit will naturally stretch over a longer period.

Planning Your Curriculum The amount of time scheduled for mathematics work will determine how much of the kindergarten *Investigations* curriculum a teacher is able to cover in the school year. You may have to make some choices as you adapt the units to your particular schedule. What is most important is finding a way to involve students in mathematics every day of the school year.

Each unit will be handled somewhat differently by every teacher. You need to be active in determining an appropriate pace and the best transition points for your class. As you read an investigation, make some preliminary decisions about how many days you will need to present the activities, based on what you know about your students and about your schedule. You may need to modify your initial plans as you proceed, and you may want to make notes in the margins of the pages as reminders for the next time you use the unit.

Help for You, the Teacher

Because we believe strongly that a new curriculum must help teachers think in new ways about mathematics and about their students' mathematical thinking processes, we have included a great deal of material to help you learn more about both.

About the Mathematics in This Unit This introductory section summarizes the essential information about the mathematics you will be teaching. It describes the unit's central mathematical ideas and the ways students will encounter them through the unit's activities.

Teacher Notes These reference notes provide practical information about the mathematics you are teaching and about our experience with how students learn. Many of the notes were written in response to actual questions from teachers or to discuss important things we saw happening in the field-test classrooms. Some teachers like to read them all before starting the unit, then review them as they come up in particular investigations.

In the kindergarten units, Teacher Notes headed "From the Classroom" contain anecdotal reflections of teachers. Some focus on classroom management issues, while others are observations of students at work. These notes offer another perspective on how an activity might unfold or how kindergarten students might become engaged with a particular material or activity.

A few Teacher Notes touch on fundamental principles of using *Investigations* and focus on the pedagogy of the kindergarten classroom:

- About Choice Time
- Materials as Tools for Learning
- Encouraging Students to Think, Reason, and Share Ideas
- Games: The Importance of Playing More Than Once

After their initial appearance, these are repeated in the back of each unit. Reviewing these notes periodically can help you reflect on important aspects of the *Investigations* curriculum.

Dialogue Boxes Sample dialogues demonstrate how students typically express their mathematical ideas, what issues and confusions arise in their thinking, and how some teachers have guided class discussions.

Many of these dialogues are word-for-word transcriptions of recorded class discussions. They are not always easy reading; sometimes it may take some effort to unravel what the students are trying to say. But this is the value of these dialogues; they offer good clues to how your students may develop and express their approaches and strategies, helping you prepare for your own class discussions.

Where to Start You may not have time to read everything the first time you use this unit. As a first-time user, you will likely focus on understanding the activities and working them out with your students. You will also want to read the few sections listed in the Contents under the heading Where to Start.

TECHNOLOGY IN THE CURRICULUM

The *Investigations* curriculum incorporates the use of two forms of technology in the classroom: calculators and computers. Calculators are assumed to be standard classroom materials, available for student use in any unit. Computers are explicitly linked to one or more units at each grade level; they are used with the unit on 2-D geometry at each grade, as well as with some of the units on measuring, data, and changes.

Using Calculators

In this curriculum, calculators are considered tools for doing mathematics, similar to pattern blocks or interlocking cubes. Just as with other tools, students must learn both *how* to use calculators correctly and *when* they are appropriate to use. This knowledge is crucial for daily life, as calculators are now a standard way of handling numerical operations, both at work and at home. Calculators are formally introduced in the grade 1 curriculum, but if available, can be introduced to kindergartners informally.

Using a calculator correctly is not a simple task; it depends on a good knowledge of the four operations and of the number system, so that students can select suitable calculations and also determine what a reasonable result would be. These skills are the basis of any work with numbers, whether or not a calculator is involved.

Unfortunately, calculators are often seen as tools to check computations with, as if other methods are somehow more fallible. Students need to understand that any computational method can be used to check any other; it's just as easy to make a mistake on the calculator as it is to make a mistake on paper or with mental arithmetic. Throughout this curriculum, we encourage students to solve computation problems in more than one way in order to double-check their accuracy. We present mental arithmetic, paper-and-pencil computation, and calculators as three possible approaches.

In this curriculum we also recognize that, despite their importance, calculators are not always appropriate in mathematics instruction. Like any tools, calculators are useful for some tasks but not for others. You will need to make decisions about when to allow students access to calculators and when to ask that they solve problems without them, so that they can concentrate on other tools and skills. At times when calculators are or are not appropriate for a particular activity, we make specific recommendations. Help your students develop their own sense of which problems they can tackle with their own reasoning and which ones might be better solved with a combination of their own reasoning and the calculator.

Managing calculators in your classroom so that they are a tool, and not a distraction, requires some planning. When calculators are first introduced, students often want to use them for everything, even problems that can be solved quite simply by other methods. However, once the novelty wears off, students are just as interested in developing their own strategies, especially when these strategies are emphasized and valued in the classroom. Over time, students will come to recognize the ease and value of solving problems mentally, with paper and pencil, or with manipulatives, while also understanding the power of the calculator to facilitate work with larger numbers.

Experience shows that if calculators are available only occasionally, students become excited and distracted when permitted to use them. They focus on the tool rather than on the mathematics. In order to learn when calculators are appropriate and when they are not, students must have easy access to them and use them routinely in their work.

If you have a calculator for each student, and if you think your students can accept the responsibility, you might allow them to keep their calculators with the rest of their individual materials, at least for the first few weeks of school. Alternatively, you might store them in boxes on a shelf, number each calculator, and assign a corresponding number to each student. This system can give students a sense of ownership while also helping you keep track of the calculators.

Using Computers

Students can use computers to approach and visualize mathematical situations in new ways. The computer allows students to construct and manipulate geometric shapes, see objects move according to rules they specify, and turn, flip, and repeat a pattern.

This curriculum calls for computers in units where they are a particularly effective tool for learning mathematics content. One unit on 2-D geometry at each of the grades 3–5 includes a core of activities that rely on access to computers, either in the classroom or in a lab. Other units on geometry, measurement, data, and changes include computer activities, but can be taught without them. In these units, however, students' experience is greatly enhanced by computer use.

The following list outlines the recommended use of computers in this curriculum:

Kindergarten
Unit: *Making Shapes and Building Blocks* (Exploring Geometry)
Software: *Shapes*
Source: provided with the unit

Grade 1
Unit: *Survey Questions and Secret Rules* (Collecting and Sorting Data)
Software: *Tabletop, Jr.*
Source: Broderbund

Unit: *Quilt Squares and Block Towns* (2-D and 3-D Geometry)
Software: *Shapes*
Source: provided with the unit

Grade 2
Unit: *Mathematical Thinking at Grade 2* (Introduction)
Software: *Shapes*
Source: provided with the unit

Unit: *Shapes, Halves, and Symmetry* (Geometry and Fractions)
Software: *Shapes*
Source: provided with the unit

Unit: *How Long? How Far?* (Measuring)
Software: *Geo-Logo*
Source: provided with the unit

Grade 3
Unit: *Flips, Turns, and Area* (2-D Geometry)
Software: *Tumbling Tetrominoes*
Source: provided with the unit

Unit: *Turtle Paths* (2-D Geometry)
Software: *Geo-Logo*
Source: provided with the unit

Grade 4
Unit: *Sunken Ships and Grid Patterns* (2-D Geometry)
Software: *Geo-Logo*
Source: provided with the unit

Grade 5
Unit: *Picturing Polygons* (2-D Geometry)
Software: *Geo-Logo*
Source: provided with the unit

Unit: *Patterns of Change* (Tables and Graphs)
Software: *Trips*
Source: provided with the unit

Unit: *Data: Kids, Cats, and Ads* (Statistics)
Software: *Tabletop, Sr.*
Source: Broderbund

The software provided with the *Investigations* units uses the power of the computer to help students explore mathematical ideas and relationships that cannot be explored in the same way with physical materials. With the *Shapes* (grades K–2) and *Tumbling Tetrominoes* (grade 3) software, students explore symmetry, pattern, rotation and reflection, area, and characteristics of 2-D shapes. With the *Geo-Logo* software (grades 2–5), students investigate rotations and reflections, coordinate geometry, the properties of 2-D shapes, and angles. The *Trips* software (grade 5) is a mathematical exploration of motion in which students run experiments and interpret data presented in graphs and tables.

We suggest that students work in pairs on the computer; this not only maximizes computer resources but also encourages students to consult, monitor, and teach one another. However, asking more than two students to work at the same computer is less effective. Managing access to computers is an issue for every classroom. The curriculum gives you explicit support for setting up a system. The units are structured on the assumption that you have enough computers for half your students to work on the machines in pairs at one time. If you do not have access to that many computers, suggestions are made for structuring class time to use the unit with fewer than five.

ABOUT ASSESSMENT

Assessment plays a critical role in teaching and learning, and it is an integral part of the *Investigations* curriculum. For a teacher using these units, assessment is an ongoing process. You observe students' discussions and explanations of their ideas and strategies on a daily basis and examine their work as it evolves. While students are busy working with materials, playing mathematical games, sharing ideas with partners, and working on projects, you have many opportunities to observe their mathematical thinking. What you learn through observation guides your decisions about how to proceed, both with the curriculum and with individual students.

Our experiences with young children suggest that they know, can explain, and can demonstrate with materials a lot more than they can represent on paper. This is one reason why it is so important to engage children in conversation, helping them explain their thinking about a problem they are solving. It is also why, in kindergarten, assessment is based exclusively on a teacher's observations of students as they work.

The way you observe students will vary throughout the year. At times you may be interested in particular strategies that students are developing to solve problems. Other times, you might want to observe how students use or do not use materials for solving problems. You may want to focus on how students interact when working in pairs or groups. You may be interested in noting the strategy that a student uses when playing a game during Choice Time. Or you may take note of student ideas and thinking during class discussions.

Assessment Tools in the Unit

Virtually every activity in the kindergarten units of the *Investigations* curriculum includes a section called Observing the Students. This section is a teacher's primary assessment tool. It offers guidelines on what to look for as students encounter the mathematics of the activity. It may suggest questions you can ask to uncover student thinking or to stimulate further investigation. When useful, a range of potential responses or examples of typical student approaches are given, along with ways to adapt the activity for students in need of more or less challenge.

Supplementing this main assessment tool in each unit are the Teacher Notes and Dialogue Boxes that contain examples of student work, teacher observations, and student conversations from real kindergarten classrooms. These resources can help you interpret experiences from your own classroom as you progress through a unit.

Documentation of Student Growth

You will probably need to develop some sort of system to record and keep track of your observations. A single observation is like a snapshot of a student's experience with a particular activity, but when considered over time, a collection of these snapshots provides an informative and detailed picture of a student. Such observations are useful in documenting and assessing student's growth, as well as in planning curriculum.

Observation Notes A few ideas that teachers have found successful for record keeping are suggested here. The most important consideration is finding a system that really works for you. All too often, keeping observation notes on a class of 20–30 students is overwhelming and time-consuming. Your goal is to find a system that is neither.

Some teachers find that a class list of names is convenient for jotting down their observations. Since the space is limited, it is not possible to write lengthy notes; however, over time, these short observations provide important information.

Other teachers keep a card file or a loose-leaf notebook with a page for each student. When something about a student's thinking strikes them as important, they jot down brief notes and the date.

Some teachers use self-sticking address labels, kept on clipboards around the classroom. After taking notes on individual students, they simply peel off each label and stick it into the appropriate student file or notebook page.

You may find that writing notes at the end of each week works well for you. For some teachers, this process helps them reflect on individual students, on the curriculum, and on the class as a whole. Planning for the next weeks' activities often grows out of these weekly reflections.

Student Portfolios Collecting samples of student work from each unit in a portfolio is another way to document a student's experience that supports your observation notes. In kindergarten, samples of student work may include constructions, patterns, or designs that students have recorded, score sheets from games they have played, and early attempts to record their problem-solving strategies on paper, using pictures, numbers, or words.

The ability to record and represent one's ideas and strategies on paper develops over time. Not all 5- and 6-year-olds will be ready for this. Even when students are ready, what they record will have meaning for them only in the moment—as they work on the activity and make their representation. You can augment this by taking dictation of a student's idea or strategy. This not only helps both you and the student recall the idea, but also gives students a model of how their ideas could be recorded on paper.

Over the school year, student work samples combined with anecdotal observations are valuable resources when you are preparing for family conferences or writing student reports. They help you communicate student growth and progress, both to families and to the students' subsequent teachers.

Assessment Overview

There are two places to turn for a preview of the assessment information in each kindergarten *Investigations* unit. The Assessment Resources column in the Unit Overview chart locates the Observing the Students section for each activity, plus any Teacher Notes and Dialogue Boxes that explain what to look for and what types of responses you might see in your classroom. Additionally, the section called About the Assessment in This Unit gives you a detailed list of questions, keyed to the mathematical emphases for each investigation, to help you observe and assess student growth. This section also includes suggestions for choosing student work to save from the unit.

These examples illustrate record keeping systems used by two different teachers for the kindergarten unit *Collecting, Counting, and Measuring*, one using the class list and the other using individual note cards to record student progress.

Emma Ruiz

3/19	Counting Jar: counts 9 balls accurately and makes another set of 9 cubes
3/24	Today's Question: compares data, "13 is 4 more than 9 because the 13 tower is 4 names taller."
4/1	Draws Counting Book pictures for 1-6, then adds pgs 7,8,9,10,11 on her own

---

Unit: Collecting, Counting, and Measuring
Activity: Inventory Bags
Date: 10/12 and 10/13

Alexa • counting sequence to 50↑ • counts 1:1 up to 12 • counts 4 bags accurately	**Luke** • counts to 30, misses 19, 20 and 29, 30 • counts by moving objects; 1:1 to 10 objects • draws circles for buttons
Ayesha • works with Oscar • counts to 15 accurately — trouble beyond 15 but Oscar helps ★ Meet to check counting • 1:1 to 8 objects?	**Maddy** • difficult to tell how much M. counted herself + how much was done by partner. Work w/ her to see.
Brendan absent 10/12, 10/13	**Miyuki** • counts aloud beyond 30 but leaves out 14 • counts 1:1 up to 10 but doesn't organize objects
Carlo • counts objects with difficulty. • remove items from bag so he works with 10 • says numbers to 10, counts objects to 6	**Oscar** • works with Ayesha • counts rotely to 20, maybe higher • double-checks his count every time — is accurate
Charlotte • completed inventory task easily without help • counts accurately up to 20 objects • represents with numbers	**Ravi** • worked w/ his aide to complete task • counts 1:1 to 5 objects • difficulty representing quantity w/ pictures
Felipe • worked well with Tarik • counted ___ — 21 in all	**Renata**

About Assessment ■ **I-9**

UNIT OVERVIEW

Making Shapes and Building Blocks

Content of This Unit In this introduction to geometry, students look for 2-D and 3-D shapes in their classroom environment. Using materials such as pattern blocks, Geoblocks, clay, and the *Shapes* software, students observe, describe, construct, and represent a variety of 2-D and 3-D shapes. They explore how shapes can be combined to make other shapes as they work on Pattern Block Puzzles and the activity Build a Block, and as they play the game Fill the Hexagons. In addition, students begin to work with 2-D representations of 3-D objects as they hunt for 3-D shapes in the environment that correspond to pictured shapes, and as they try to match particular Geoblocks to 2-D outlines of the block faces.

In this unit, the use of computer activities is optional. However, if you have computers available, we strongly recommend that you use the software. The computer work is integrated into the unit and enriches the work students do.

Connections with Other Units If you are doing the full-year *Investigations* curriculum, this is the fifth of six units. Pattern blocks and Geoblocks were introduced in the first unit, *Mathematical Thinking in Kindergarten;* now students use these tools to explore geometric concepts. Students' work in this unit with sorting and classifying geometric shapes connects to explorations they began in the data unit, *Counting Ourselves and Others.*

This unit can also be used successfully at grade 1, depending on the needs and previous experience of your students.

Investigations Curriculum ■ Suggested Kindergarten Sequence

Mathematical Thinking in Kindergarten (Introduction)

Pattern Trains and Hopscotch Paths (Exploring Pattern)

Collecting, Counting, and Measuring (Developing Number Sense)

Counting Ourselves and Others (Exploring Data)

▶ *Making Shapes and Building Blocks* (Exploring Geometry)

How Many in All? (Counting and the Number System)

Investigation 1 ■ 2-D Shapes Around Us

Class Sessions	Activities	Pacing
FOCUS TIME (p. 4) Looking at 2-D Shapes	A Book About Shapes Planning a Class Book of Shapes Homework: Family Connection Extension: Potato-Cut Printing	1 session
CHOICE TIME (p. 12)	Book of Shapes Pattern Block Pictures Shape Mural	3–4 sessions
Classroom Routines	Attendance and Calendar (daily) Counting Jar, Today's Question, and Patterns on the Pocket Chart (weekly or as appropriate)	

Mathematical Emphasis

- Recognizing shapes in the environment
- Observing and describing two-dimensional (2-D) shapes
- Developing vocabulary to describe 2-D shapes
- Becoming familiar with the names of 2-D shapes
- Relating 2-D shapes to real-world objects

Assessment Resources

Observing the Students:

- Book of Shapes (pp. 9 and 13)
- Pattern Block Pictures (p. 16)
- Shape Mural (p. 19)

Materials

The Shape of Things or a similar book

Large paper (11 by 17 inches)

Pattern blocks

Butcher paper (or other mural paper)

Paper pattern blocks

Crayons, colored pencils, or markers

Art supplies (stencils, rubber stamps, stickers)

Glue or paste

Stapler or hole-punch and string (optional)

Scoops and trays or cardboard (optional)

Teaching resource sheets

Pattern blocks

Unit Overview ■ I-11

Investigation 2 ▪ Exploring Shapes with the Computer

Class Sessions	Activities	Pacing
FOCUS TIME (p. 28) Introducing the *Shapes* Software	Introducing the Computer Sharing Work in Free Explore	1 session*
CHOICE TIME (p. 31)	Free Explore with *Shapes* on the Computer Pattern Block Puzzles Book of Shapes Shape Mural	3–4 sessions*
Classroom Routines	Attendance and Calendar (daily) Counting Jar, Today's Question, and Patterns on the Pocket Chart (weekly or as appropriate)	

* If you do not intend to use computers with this unit, plan to skip Investigation 2 and introduce the Pattern Block Puzzles in Investigation 3.

Mathematical Emphasis

- Exploring the *Shapes* software
- Working with a partner on the computer
- Visualizing how to move a shape so that it is oriented correctly to fit into a design
- Finding combinations of shapes that fill an area
- Building knowledge about the relationships among pattern block shapes
- Developing vocabulary to describe 2-D shapes
- Relating 2-D shapes to real-world objects

Assessment Resources

Observing the Students:

- Free Explore with *Shapes* on the Computer (p. 33)
- Pattern Block Puzzles (p. 35)

Materials

Computers with *Shapes* software installed

Pattern blocks

Paper pattern blocks (optional)

Teaching resource sheets

Shape mural, with drawing and art supplies, from Investigation 1

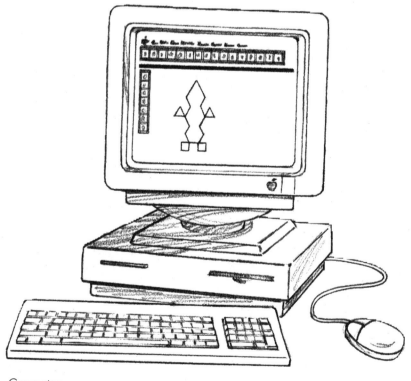

Computer

I-12 ▪ *Making Shapes and Building Blocks*

Investigation 3 ■ Looking at 3-D Shapes

Class Sessions	Activities	Pacing
FOCUS TIME (p. 40) 3-D Shapes in the Classroom	Looking at 3-D Shapes Shape Hunt Comparing 3-D Shapes Homework: Looking for Shapes at Home Extension: Comparing Geometric Solids	1–2 sessions
CHOICE TIME (p. 46)	Shape Hunt Exploring Geoblocks The Shape of Things on the Computer Pattern Block Puzzles	4–5 sessions
Classroom Routines	Attendance and Calendar (daily) Counting Jar, Today's Question, and Patterns on the Pocket Chart (weekly or as appropriate)	

Mathematical Emphasis

- Recognizing shapes in the environment
- Developing vocabulary to describe 2-D and 3-D shapes
- Becoming familiar with the names of 2-D and 3-D shapes
- Relating a 3-D object to a 2-D picture of its geometric shape
- Exploring and using the *Shapes* software
- Finding combinations of shapes that fill an area
- Building knowledge about the relationships among pattern block shapes
- Picturing the shape that will fit a particular space or design
- Visualizing how a shape needs to be moved or turned in order to fit into a particular space or design

Assessment Resources

Observing the Students:

- Shape Hunt (pp. 43 and 47)
- Exploring Geoblocks (p. 49)
- The Shape of Things on the Computer (p. 52)

Dialogue Box: It Looks Like a Ball (p. 54)

Dialogue Box: Solving Computer Puzzles (p. 58)

Materials

Geometric solids (set of six)

Geoblocks

Shoe boxes or other containers

Chart paper (optional)

Clipboards—cardboard and paper clips (optional)

Computers with *Shapes* installed

Pattern blocks

Paper pattern blocks (optional)

Glue sticks or paste (optional)

Student Sheet 1

Teaching resource sheets

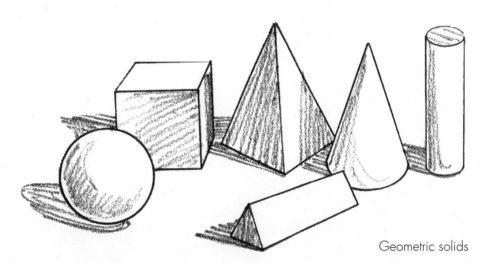

Geometric solids

Investigation 4 ■ Making Shapes and Building Blocks

Class Sessions	Activities	Pacing
FOCUS TIME (p. 62) Clay Shapes	Describing Shapes Making Shapes with Clay Sharing Our Clay Shapes Extension: Rope Shapes	1–2 sessions
CHOICE TIME (p. 68)	Clay Shapes Fill the Hexagons Build a Block Quick Images on the Computer Pattern Block Puzzles	4–5 sessions
Classroom Routines	Attendance and Calendar (daily) Counting Jar, Today's Question, and Patterns on the Pocket Chart (weekly or as appropriate)	

Mathematical Emphasis

- Describing and becoming familiar with the attributes of 2-D shapes
- Constructing 2-D shapes
- Finding combinations of shapes that fill an area
- Building knowledge about the relationships among pattern block shapes
- Combining smaller 3-D shapes to make a larger 3-D shape
- Analyzing visual images
- Describing position of and spatial relationships among objects

Assessment Resources

Observing the Students:

- Clay Shapes (pp. 65 and 69)
- Fill the Hexagons (p. 72)
- Build a Block (p. 74)
- Quick Images on the Computer (p. 78)

Teacher Note: Making Shapes (p. 79)

Dialogue Box: Three Pointy Corners (p. 80)

Dialogue Box: Circles and Ovals (p. 81)

Materials

Clay or playdough
Cardboard mats
Pattern blocks
Blank 1-inch cubes
Stick-on labels
Geoblocks
Computers with *Shapes* installed
Small cups
Tray or cardboard, and paper or cloth to conceal it
Paper pattern blocks (optional)
Glue sticks or paste (optional)
Student Sheet 2
Teaching resource sheets

Clay or playdough

Investigation 5 ■ 2-D Faces on 3-D Blocks

Class Sessions	Activities	Pacing
FOCUS TIME (p. 84) A Close Look at Geoblocks	A Close Look at Geoblocks	1 session
CHOICE TIME (p. 88)	Matching Faces Geoblock Match-Up Planning Pictures on the Computer Clay Shapes Fill the Hexagons	4–5 sessions
Classroom Routines	Attendance and Calendar (daily) Counting Jar, Today's Question, and Patterns on the Pocket Chart (weekly or as appropriate)	

Mathematical Emphasis

- Observing and describing attributes of 3-D shapes
- Looking at 3-D objects as wholes and as having parts
- Observing similarities and differences between the faces of different 3-D shapes
- Matching a 3-D block to a 2-D outline of one of the block faces

Assessment Resources

Observing the Students:

- Matching Faces (p. 89)
- Geoblock Match-Up (p. 91)
- Planning Pictures on the Computer (p. 94)

Materials

Geoblocks

Computers with *Shapes* installed

Clay or playdough

Cardboard mats

Pattern blocks

Pattern block game cubes (prepared for Investigation 4)

Teaching resource sheets

Geoblocks

MATERIALS LIST

Following are the basic materials needed for the activities in this unit. Many items can be purchased from the publisher, either individually or in the Teacher Resource Package and the Student Materials Kit for kindergarten. Detailed information is available on the *Investigations* order form. To obtain this form, call toll-free 1-800-872-1100 and ask for a Dale Seymour customer service representative.

Pattern blocks: 1 bucket per 4–6 students

Set of six geometric solids (cube, cylinder, sphere, triangular prism, cone, pyramid)

Geoblocks: 2 sets per classroom, divided into smaller sets

Clay or playdough: one 3-inch ball per student

Paper pattern blocks: 1 manufactured set, or prepare from masters on pp. 160–165

1-inch blank cubes with stick-on labels for making pattern block game cubes

Computers: Macintosh II or above, with 4 MB of internal memory (RAM) and Apple System Software 7.0 or later

Apple Macintosh disk, *Shapes—Making Shapes* (packaged with this book)

The Shape of Things by Dayle Ann Dodds (Candlewick Press, 1994) or a similar book about shapes in the environment

Small cups: 1 per student

Small tray, and paper or cloth large enough to cover it completely

Large paper (11 by 17 inches): 1 sheet per student, plus extras

Butcher paper, oak tag, or other large paper for a mural

Chart paper (optional)

Sturdy cardboard for mats

"Clipboards" made from a piece of stiff cardboard and a paper clip: 1 per student (optional)

Crayons, markers, and/or colored pencils

Art supplies, such as stencils, rubber stamps, and stickers

Glue sticks or paste

A stapler or hole-punch and string, for binding a class book (optional)

Shoe boxes or other containers for storing smaller Geoblock sets

The following materials are provided at the end of the unit as blackline masters:

Family Letter (p. 156)

Student Sheets 1–2 (p. 176)

Teaching Resources:

    Shape Cutouts A–F (p. 157)

    Pattern Block Cutouts (p. 160)

    Pattern Block Puzzles 1–10 (p. 166)

    Shape Hunt at Home (p. 177)

    Make-a-Shape Cards (p. 178)

    Geoblock Match-Up Gameboards (p. 182)

    Choice Board Art (p. 188)

Related Children's Literature

Angelou, Maya. *My Painted House, My Friendly Chicken and Me.* New York: Clarkson Potter, 1994.

Burns, Marilyn. *The Greedy Triangle.* New York: Scholastic, 1994

Dodds, Dayle Ann. *The Shape of Things.* Boston: Candlewick Press, 1994.

Ehlert, Lois. *Color Zoo.* New York: HarperCollins, 1989.

Falwell, Cathryn. *Shape Space.* New York: William Morrow, 1992

Hoban, Tana. *Shapes, Shapes, Shapes.* New York: Greenwillow Books, 1986.

Kightley, Rosalind. *Shapes.* New York: Little, Brown, 1986.

Rogers, Paul. *The Shapes Game.* New York: Henry Holt, 1989.

Testa, Fulvio. *If You Look Around You.* New York: Dial Books, 1983

ABOUT THE MATHEMATICS IN THIS UNIT

Students entering kindergarten bring with them a good deal of informal experience with geometry. As young children use their eyes and hands to interact with shapes and images in their everyday world, they are developing an intuitive sense of how those shapes and images are the same and how they are different. They may not have the mathematical language to describe important attributes such as *curves*, *angles*, and *straight sides*, but most can say which objects are "round," which are "pointy," and which are "flat." This geometry unit, *Making Shapes and Building Blocks*, builds on students' firsthand knowledge of shapes to further develop their spatial sense and deepen their understanding of the two- and three-dimensional world in which they live.

As students begin to identify the different shapes that make up the world, they are encouraged to use their own words to tell what those shapes, both 2-D and 3-D, look like. Young children begin to form an image of a shape through its association with familiar objects. For example, they might describe a triangle by relating it to other triangle-like shapes with which they are familiar, such as a folded paper hat or the roof of a house. Through these associations, they are forming a mental image of a triangle. With further exploration, students begin to make generalizations about all three objects—the triangle, the paper hat, and the roof—as they notice the shared attributes: three straight sides and three angles (or "points," as most young students describe them).

Many activities in this unit emphasize observing and describing shapes. Others present an experience that is related and yet somewhat different: *constructing* shapes. Students manipulate ropes of clay to make the outlines of triangles, squares, circles, and other 2-D shapes. This brings a tactile experience to the exploration of geometry. Working with a rope of clay, students get a hands-on sense of straight sides and angles, and how these parts combine to form a triangle. In order to construct 2-D shapes, students must integrate a mental or visual image of the shape with the tactile experience of forming the shape. This work helps to deepen students' understanding of just what constitutes a particular shape—for example, what makes a triangle a triangle—and how it compares to other shapes.

Throughout this unit students use pattern blocks, Geoblocks, and the *Shapes* software to explore another geometric idea: that shapes can be combined or subdivided to make other shapes. For example, in the game Fill the Hexagons, students discover a variety of ways that triangles, trapezoids, and rhombuses can be combined to form a hexagon shape. In the Build a Block activity, they investigate how two or more smaller 3-D shapes can be combined to form a larger rectangular solid. By putting shapes together and taking shapes apart, students deepen their understanding of how these shapes are related.

At the beginning of each investigation, the Mathematical Emphasis section tells you what is most important for students to learn about during that investigation. Many of these mathematical understandings and processes are difficult and complex. Students gradually learn more and more about each idea over many years of schooling. Individual students will begin and end the unit with different levels of knowledge and skill, but all will learn more about geometric shapes, their relationships, and their properties.

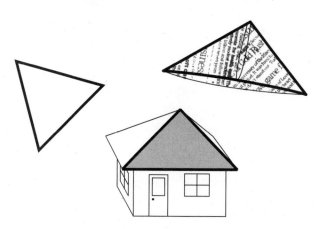

A triangle may suggest to young students the shape of a hat or a roof.

ABOUT THE ASSESSMENT IN THIS UNIT

Throughout the *Investigations* curriculum, there are many opportunities for ongoing daily assessment as you observe, listen to, and interact with students at work. You can use almost any activity in this unit to assess your students' needs and strengths. Listed below are questions to help you focus your observations in each investigation. You may want to keep track of your observations for each student to help you plan your curriculum and monitor students' growth. Suggestions for documenting student growth can be found in the section About Assessment (p. I-8).

Investigation 1: 2-D Shapes Around Us

- How do students talk about and describe two-dimensional shapes? Do they use names? informal language? size? attributes of the shapes?

- In making a shape picture, how do students decide which shapes to use? Do they start by thinking of a 3-D object? ("My house is like a box, so I'm going to use a rectangle.") Or do they start with a 2-D shape? ("This circle looks like a sun or a ball.")

- Do students combine pattern blocks to make other shapes? Do they have a sense that combinations of blocks can be substituted for other blocks?

Investigation 2: Exploring Shapes with the Computer

- In the *Shapes* software, which tools are students using and experimenting with? How comfortable are they using these tools? Do they have a sense of what will happen when they choose a particular tool?

- What kinds of patterns, designs, or pictures do students create? How do they decide which shapes to select? Do they choose them randomly? Do they begin to plan what shapes they will need for their design and select these deliberately?

- Can students move a shape into the right position and orientation on the screen?

- Do you see evidence that students know how to make the same shape in different ways? Can students use equivalents among the shapes to help them change their designs?

Investigation 3: Looking at 3-D Shapes

- How do students talk about and describe three-dimensional shapes? Do they refer to 3-D shapes using 2-D names? Are students able to relate a 3-D object to its two-dimensional picture?

- How do they describe the Geoblocks? Do they describe differences in size? Do they notice the 2-D shapes on the Geoblock faces? What words do they use to talk about these shapes? Do students have a sense that pairs or combinations of blocks can be substituted for other blocks?

- Are students becoming more comfortable with the *Shapes* software? Are they using the tools that you have introduced?

- While solving pattern block puzzles, how do students decide what shapes to select? Do they select shapes randomly? deliberately?

- Do students realize there are many ways to solve a puzzle? Are they demonstrating any knowledge of the equivalence between pattern block shapes?

Investigation 4: Making Shapes and Building Blocks

- Are students able to make a variety of shapes with clay ropes? Do they consult the shape posters or shape cards or do they seem to "just know" the shapes they want to construct?

- How do students make their shapes? Do they seem to know specific attributes of a shape, such as "a triangle has three sides"? Do they make individual sides and then attach them at the corners, or do they use one long clay rope to outline a shape?

- How fluently can students place pattern blocks to fill hexagon outlines? Do they plan ahead? What relationships among the shapes do students use in deciding how to place their blocks? Is there evidence that they are beginning to see and use equivalents among the shapes?

- How do students put Geoblocks together to build a larger block? Do they randomly choose blocks, or do they look specifically for two blocks that would be half of a larger block? Do they use smaller cubes to build larger cubes?

- What strategies do students use for reconstructing a design in Quick Images? Do they first retrieve the blocks seen, then think about how to arrange them? Do they try to remember the blocks in a particular order? Do they associate the image with a real-world object to help them recreate it?
- Which tools are students using in *Shapes* to build copies of the images? Do they have a sense of what will happen when they choose a particular tool?

Investigation 5: 2-D Faces on 3-D Blocks
- Do students find Geoblock faces to match other block faces and outlines easily or with difficulty?
- Do students keep in mind all the faces on one block (or all the outlines on their gameboard) as they work? For example, when a shape doesn't match one face of another block (or one particular outline), do they check to see if it matches any of the others?
- Do students notice that the triangular prisms have some rectangular faces?
- Can students differentiate between different sizes of the same shape?
- What vocabulary do students use to describe the 3-D shapes? the 2-D faces? Do they use language that describes differences between shapes? For example, do they talk about shapes as being *thick* or *thin, tall* or *short*?
- When planning pictures with pattern blocks, how comfortable are students selecting, moving, turning, and flipping pattern blocks, both on and off the computer? On the computer, do students accurately choose blocks and then motions to match the picture that they are copying?
- Do students recognize when there is a discrepancy between their original picture and their copy? How do they handle this kind of discrepancy? Do they adjust their computer design? their pattern block design?

Choosing Student Work to Save
As the unit ends, you might use one if the following options for creating a record of students' work.

- Look through any work samples that students have created during the unit and select a few examples to save in a portfolio or to share with parents during parent conferences. For this unit, you might include the student's page from the class Book of Shapes, pattern block designs that the student has recorded on paper, or any printouts of the student's work in *Shapes*.
- Look back through your observation notes for each student. Depending on how you have organized this information, you might select a sample of observations that document student growth and understanding, or you may want to write a brief paragraph summarizing each student's work during the unit.
- Together with your students, write a group letter to families that describes the mathematical work students did during the unit. They might tell which activities they enjoyed most, what important things they discovered about 2-D and 3-D shapes, and what they think they learned from the work they did. In one classroom, the teacher prepared the students' letter on the computer and also included a few samples of work. Every child in the class signed the letter before it was copied and distributed to families.

PREVIEW FOR THE LINGUISTICALLY DIVERSE CLASSROOM

In the *Investigations* curriculum, mathematical vocabulary is introduced naturally during the activities. We don't ask students to learn definitions of new terms; rather, they come to understand such words as *triangle, add, compare, data,* and *graph* by hearing them used frequently in discussion as they investigate new concepts. This approach is compatible with current theories of second-language acquisition, which emphasize the use of new vocabulary in meaningful contexts while students are actively involved with objects, pictures, and physical movement.

Listed below are some key words used in this unit that will not be new to most English speakers at this age level, but may be unfamiliar to students with limited English proficiency. You will want to spend additional time working on these words with your students who are learning English. If your students are working with a second-language teacher, you might enlist your colleague's aid in familiarizing students with these words, before and during this unit. In the classroom, look for opportunities for students to hear and use these words. Activities you can use to present the words are given in the appendix, Vocabulary Support for Second-Language Learners.

outline, fill in Students fill in the outlines of Pattern Block Puzzles; they make the outlines of shapes with ropes of clay; and they match Geoblock faces to outlines on a gameboard.

names of everyday objects Throughout this unit, students look for similarities between 2-D and 3-D shapes and everyday objects in the world around them. They hunt for such objects in the classroom, and they represent such objects in the Book of Shapes and on a class Shape Mural. While you cannot always anticipate the words that will arise in these contexts, remember to point to particular objects being discussed and draw simple sketches or diagrams on the board to illustrate key words, in order to keep the discussion comprehensible.

Multicultural Extensions for All Students

Whenever possible, encourage students to share words, objects, customs, or any aspects of daily life from their own cultures and backgrounds that are relevant to the activities in this unit. For example:

- Many ethnic groups have distinctive woven fabrics that show an interesting use of two-dimensional shapes. Present any examples of these that you or families of your students can supply and point out the shapes.

- A variation to the Shape Mural activity in Investigation 1 suggests that students plan decorations for the side of a house, in a style similar to the mural art painted by the Ndebele women of South Africa. This art is illustrated in Maya Angelou's book *My Painted House, My Friendly Chicken and Me* (Clarkson Potter, 1994). As you read the book to the class, focus attention on the mural designs that show in many of the photographs.

- In Investigation 4, when you prepare the four shape posters as models for the students' clay shapes, include the terms for *square, circle, triangle,* and *rectangle* in the languages of students in your class.

Investigations

INVESTIGATION 1

2-D Shapes Around Us

Focus Time

Looking at 2-D Shapes (p. 4)
Students begin to think about two-dimensional shapes as they look for and talk about the shapes around their classroom. Together they look at a book about the shape of things in the world and then design individual pages for a class Book of Shapes.

Choice Time

Book of Shapes (p. 12)
Students continue to work on their pages for the class Book of Shapes.

Pattern Block Pictures (p. 14)
After exploring pattern blocks, students make a permanent representation of their pattern block designs or pictures.

Shape Mural (p. 17)
Students work together to design and make a shape mural, using cutout paper shapes to represent objects and people.

A Note on Using Computers

Shapes, a software program that enables students to work with 2-D shapes on the computer, is introduced in Investigation 2. The **Teacher Note**, Managing the Computer Activities (p. 36), discusses options for integrating computer work with this unit. If you will be using computers, you may want to merge Investigations 1 and 2, introducing the computer to small groups while the rest of the class works on Choice Time. If computers are not available, plan to skip Investigation 2.

Mathematical Emphasis

- Recognizing shapes in the environment
- Observing and describing two-dimensional (2-D) shapes
- Developing vocabulary to describe 2-D shapes
- Becoming familiar with the names of 2-D shapes
- Relating 2-D shapes to real-world objects

Teacher Support

Teacher Notes
Learning to See Shapes in the Environment (p. 21)
How Young Children Learn About Shapes (p. 22)
About Pattern Blocks (p. 24)

Dialogue Box
Ideas for a Book of Shapes (p. 25)

"A square can be a present."

INVESTIGATION 1

What to Plan Ahead of Time

Focus Time Materials

Looking at 2-D Shapes

- *The Shape of Things* by Dayle Ann Dodds (Candlewick Press, 1994) or a similar book about shapes in the environment (see p. 4 for alternative titles)
- Shape Cutouts A–F (pp. 157–159): about 10 of each, copied on paper of different colors
- Large paper (11 by 17 inches): 1 sheet per student, plus extras
- Crayons, markers, and/or colored pencils
- Art supplies, such as stencils, rubber stamps, and stickers
- Glue sticks or paste

Choice Time Materials

Book of Shapes

- Shape Cutouts in a variety of colors, large paper, coloring materials, art supplies, and glue sticks from Focus Time
- A stapler or hole-punch and string, for binding the class book (optional)

Pattern Block Pictures

- Pattern blocks: 1 bucket per 4–6 students
- Small cups or containers to use as scoops (optional)
- Paper pattern blocks: 1 manufactured set, or prepare from masters on pp. 160–165
- Unlined paper: 1 sheet per student
- Glue sticks

Shape Mural

- *The Shape of Things* by Dayle Ann Dodds (or similar book of shapes)
- Butcher paper, oak tag, or other large paper for a mural
- Paper pattern blocks: 1 manufactured set, or prepare from masters on pp. 160–165
- Shapes Cutouts remaining from work on the Book of Shapes (make extras as needed)
- Coloring materials and art supplies
- Glue sticks or paste
- Chart paper (optional)

Family Connection

- Family letter (p. 156) or *Investigations* at Home: 1 per family

Investigation 1: 2-D Shapes Around Us

Focus Time

Looking at 2-D Shapes

What Happens

Students look for and talk about the shapes they see in their classroom as they look at a book about the shape of things they may have seen in the world around them. As they discuss the illustrations and text, they think about how they might design a book that is similar. Each student then designs a page to contribute to a class Book of Shapes. Students' work focuses on:

- observing and describing shapes
- developing vocabulary to describe shapes
- becoming familiar with the names of shapes
- relating 2-D shapes to real-world objects
- using shapes to make a picture

Materials and Preparation

- Obtain a copy of *The Shape of Things* by Dayle Ann Dodds (Candlewick Press, 1994) or a similar book about shapes in the environment, such as *Color Zoo* by Lois Ehlert (HarperCollins, 1989), *The Shapes Game* by Paul Rogers (Holt, 1989), or *Shapes* by Rosalind Kightley (Little, Brown, 1986).
- Using paper in a variety of colors, make about 10 copies of the six Shape Cutouts A–F (pp. 157–159). Cut out the shapes and store in separate resealable plastic bags, small boxes, or other containers, each labeled with an example of the shape inside.
- To make pages for the class Book of Shapes, provide 11-by-17-inch paper (1 sheet per student, plus extras); crayons, markers, and/or colored pencils; and art supplies, such as stencils, rubber stamps, and stickers. Have available glue sticks or paste to share.

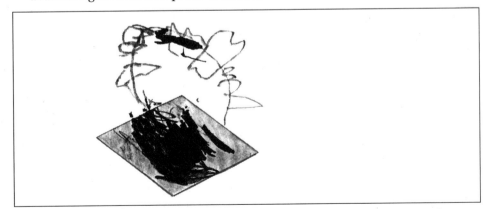

"I picked the diamond. This is a necklace."

4 ■ *Investigation 1: 2-D Shapes Around Us*

Activity

A Book About Shapes

In her book *The Shape of Things,* Dayle Ann Dodds uses such shapes as squares, circles, rectangles, triangles, and ovals to create objects in the environment: a square becomes a house; an oval becomes a hen's egg; a circle becomes a Ferris wheel. The rhythmic text (in a repeating pattern) and the colorful illustrations are engaging for kindergarten students. Present this title (or one of the similar books suggested) to introduce shapes to the whole class.

The book we're going to read today is called *The Shape of Things*. What do you think this book is going to be about? What do you notice about the cover? Does it give you any clues?

After students share their ideas, show them the inside cover and facing page, filled with colorful shapes. If you have investigated patterns in your class—perhaps with the second kindergarten unit in the year-long *Investigations* sequence, *Pattern Trains and Hopscotch Paths*—students will likely comment on the rows of linear patterns. They might read some of the patterns aloud, or talk about the shapes used to make the patterns. Notice what students mention as they read and discuss the patterns. Do they notice color? shape? size? orientation? some combination of these?

After reading the entire book, turn back and ask students to look carefully at some of the pages.

What are some of the things you noticed about the pictures and the words in this book?

Students may comment on the scene depicted on each right-hand page and the shapes used to make it. They may mention the large shape on the facing left-hand page, or the repeating pattern that borders the page, or the patterns in the phrasing of the text.

As students offer their comments and observations, listen to the words they are using to describe and name the shapes. This can provide you with some preliminary information about what they know and understand about shapes. See the **Teacher Note,** Learning to See Shapes in the Environment (p. 21), for more information about the informal knowledge and experience young children bring to the study of geometry. See also the **Teacher Note,** How Young Children Learn About Shapes (p. 22), for a discussion of using geometric terms in kindergarten.

Activity

Planning a Class Book of Shapes

If we wanted to put together our own book that was like *The Shape of Things,* what sorts of things would we want to include on each page? What was special about the pictures in this book? What does every page have?

Hold the book open and turn through the pages as students think about and respond to these questions. As students offer suggestions, use a page in the book to illustrate the idea for the class. To help students listen and respond to each other, encourage them to share a characteristic of the book that no one else has mentioned.

Gather students' ideas, perhaps recording them in a simple list on chart paper. A list might include these ideas:

- Each page has its own shape. (square, circle, triangle, rectangle, oval, diamond)
- Each shape makes part of a picture. (Two triangles make a sailboat.)
- The words for each page start the same way. ("A [shape] is just a [shape]...")
- Every page has a border of shapes, the same as the shape for that page. (Thus, the border for the square page is made up of squares.)
- The border of each page is a repeating pattern. (big rectangle, little rectangle)

Explain that as a class, students will put together a shape book similar to *The Shape of Things.* Each student will design one page, choosing one of the same shapes they saw in the book.

Show students Shape Cutouts A–F. Brainstorm some possibilities for how they might use a few of the shapes.

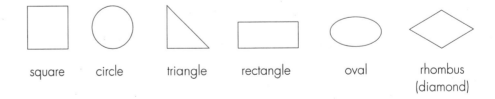

square   circle   triangle   rectangle   oval   rhombus (diamond)

Suppose you pick the square for your page. The illustrator in this book used a square to make a house. What else could you make with a square?

See the **Dialogue Box,** Ideas for a Book of Shapes, for one teacher's introduction of this activity to the class.

As needed, find a way to ensure that students select a variety of shapes, so that your class book has at least one page for each of the six shapes. Students start by gluing their selected shape onto the large paper. They can use drawing materials to add details and other elements to their shape—a roof to a house, arms and legs to a creature, a tail to a kite.

Keep *The Shape of Things* in a central location for students to use as a resource, but encourage students to design a new picture rather than to copy one directly from the book.

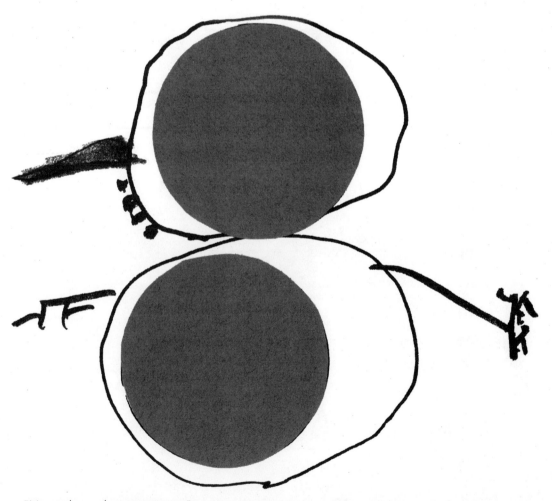

"My circles make a snowman."

Depending on the experience of your students, encourage them to write a description of their picture when they are finished, or have an adult take dictation on their page. You might want to brainstorm with the group a line of text to be repeated on each page. Then you might write a model on the board or chart paper for students to copy. For example:

A _____ is just a _____ until you add _____.

Then it is a _____!

Students can write or draw their chosen shape in the blanks to complete the text.

During the remainder of Focus Time, students can get started on their pages. Be sure they understand that this activity will continue into Choice Time on the following days so that they are not anxious about having enough time to design and finish a page within this class period.

"A triangle is just a triangle until you add a tutu. Then it is a ballerina!" This student's page includes a triangle border made from paper pattern blocks (green triangles and tan rhombuses cut in half).

"I turned the square into a diamond to make my kite."

Observing the Students

Observe students as they design and create their pictures for the class Book of Shapes. If students are having a difficult time getting started or if their ideas are limited by the illustrations in the book, enlist the help of other students to brainstorm possibilities or point out some shapes in the classroom, such as the bookshelf that looks like a rectangle.

- How do students talk about and describe shapes? Do they use names? informal language? color? size? attributes of the shapes?
- How do students decide which shapes to use in designing a particular object or scene? Do they relate the shape of the 3-D object to the 2-D shape they use? For example: "My apartment building is like a tall box, so I'm going to use a rectangle." Or do they look at a 2-D shape and assign it to an object? For example: "This circle looks like a sun or a ball."
- What attributes of shape do students notice? Do they mention the number of sides? relationship between those sides? size of the shape? orientation of the shape? What language do they use to describe these attributes?
- Do students create a pattern around the border of their page? What kind of pattern? Does the pattern include shapes?

Focus Time Follow-Up

Homework

Family Connection Send home the signed family letter or the *Investigations* at Home booklet to introduce your work in this geometry unit.

Extension

Potato-Cut Printing As students look at the borders in *The Shape of Things,* explain that they were made by potato-cut printing. Potato-cut printing or sponge printing makes a good kindergarten art project. Students might design borders for your class book, make shape pictures, or design other borders or artwork for the classroom.

Choice Time

Three Choices If Choice Time is not already a standard part of your kindergarten program, refer to the **Teacher Note,** About Choice Time (p. 96), for more information.

These three independent Choice Time activities support students' work as they continue to think about 2-D shapes in the world around them. On the first day of Choice Time, students can continue to work on their pages for the class Book of Shapes. In addition, students might begin Pattern Block Pictures (p. 14). The Shape Mural (p. 17) can be introduced on the following day as a third choice.

Note that if you plan to use computers and are merging Investigations 1 and 2, you will eventually be adding the computer activity Free Explore with *Shapes* (p. 31) and Pattern Block Puzzles (p. 34) to this list of choices.

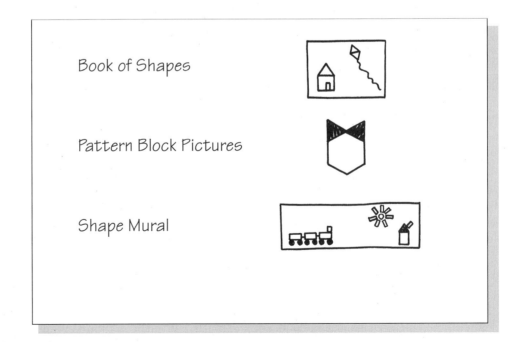

Exploring Clay In Investigation 4, students will be using clay or playdough to make shapes. If your students have not yet had the chance to freely explore this material, include Exploring Clay as an optional fourth choice here, to be repeated through the first three investigations. Use commercially available clay or playdough or make your own.

PLAYDOUGH

2 cups flour

1 cup salt

2 tablespoons cooking oil

water

food coloring

Combine ingredients using just enough water to allow mixture to be kneaded. Knead until smooth. Store in an airtight container.

Choice Time

Book of Shapes

What Happens

Students continue to work on the class Book of Shapes begun during Focus Time. Each student makes a page about a particular shape to contribute to the class book. Their work focuses on:

- observing and describing shapes
- developing vocabulary to describe shapes
- becoming familiar with the names of shapes
- relating 2-D shapes to real-world objects
- using shapes to design a picture

Materials and Preparation

- Students will need the large pages they began during Focus Time, as well as the supply of Shape Cutouts, coloring materials, and art supplies.
- Make available *The Shape of Things* (or other shape book from Focus Time) for student reference.
- Have available a stapler or a hole-punch and string for binding the class book (optional).

Activity

Students continue working on their page for the class Book of Shapes. You might gather students together before Choice Time, to remind them of their work in Focus Time and to recall as a group the things that are special or important about this book. For example:

- Each page has its own shape.
- Each shape makes part of a picture.
- The words for each page start the same way.
- Every page has a border of shapes, the same as the shape for that page.
- The border of each page is a repeating pattern.

Observing the Students

As you observe students working, ask them to tell you about what's on their page for the Book of Shapes.

- How do students talk about and describe shapes? Do they use names? informal language? color? size? attributes of the shapes?
- How do students decide which shapes to use in a designing a particular object or scene? Do they relate the shape of the 3-D object to the 2-D shape they use? ("The pond in the park is sort of round and sort of long, like this [oval].") Or do they look at a 2-D shape and assign it to an object? ("This triangle looks like a hat.")
- What attributes of shape (number of sides, relationship between those sides, size, orientation) do students comment upon? What language do they use to describe these attributes?
- Do students create a pattern around the border of their page? What kind of pattern? Does the pattern incorporate shapes?

Once students have finished their pages, bind them into a class book by stapling them together or by using a hole-punch and some string.

Variations

- Students might make a second page for the class Book of Shapes.
- Students could make their own Book of Shapes to take home, perhaps on smaller sheets of paper.

Choice Time

Pattern Block Pictures

What Happens

After free exploration of pattern blocks, students glue down paper pattern blocks to make a permanent representation of their designs or pictures. Their work focuses on:

- exploring pattern blocks and their attributes
- using informal language to describe geometric shapes
- making a 2-D representation

Materials and Preparation

- Make available your class set of pattern blocks, allowing 1 bucket per 4–6 students. You might also provide small cups or containers to use as scoops (optional).
- Provide a manufactured set of paper pattern blocks, or prepare some by copying and cutting apart the Pattern Block Cutouts (pp. 160–165). Store each shape in its own container (resealable plastic bag, small box) labeled with an example of the shape it contains. Enlist parent or classroom volunteers to help with this preparation. Your supply of paper pattern blocks will also be used for the Shape Mural (p. 17). **Note:** Students tend to use *lots* of hexagons.
- Supply unlined paper and glue sticks for students' representations.
- If you are introducing pattern blocks for the first time, read the **Teacher Notes,** About Pattern Blocks (p. 24) and Materials as Tools for Learning (p. 99), for information on the mathematics and management of these manipulatives.

Activity

Note: If this is your students' first introduction to pattern blocks or if they have had limited experience with them, it is important to offer extended opportunities to freely explore the blocks. You might include Exploring Pattern Blocks for Choice Time during the first three investigations, and then introduce Pattern Block Pictures for Choice Time during Investigation 4. If you are doing the full-year *Investigations* curriculum, your students will be familiar with pattern blocks from their work in other units and can start right in on Pattern Block Pictures.

Exploring Pattern Blocks Briefly introduce the pattern blocks to the whole group. If this material is new to the class, discuss where the blocks are stored and how they will be used and cared for.

If students are familiar with the blocks, ask how they have used them in the past and if they know the names of any of them. Remember that kindergarten students are not expected to use the formal geometric terms, as explained in the **Teacher Note,** How Young Children Learn About Shapes (p. 22).

During free exploration of the blocks, many students are able to set themselves tasks, while others need some support and structure to take their first steps. When students are hesitant, help them notice what their classmates are doing. You might suggest starting with a hexagon, or making a "wall" around a "garden," or making a design using exactly 10 pattern blocks (you can adjust the numbers for different students). Some teachers require that all designs fit on a standard sheet of paper. Using paper as a mat in this way helps students contain their work and also limits the number of blocks a single student can use.

Pattern Block Pictures Once students have had numerous opportunities to freely explore the pattern blocks, they can work on transferring their pictures or designs to paper. To introduce this activity, first make and display a simple pattern block design.

Many of you have made interesting designs and pictures with the pattern blocks. We have pattern blocks made out of paper, and you can use these to make a copy of a block design you really like.

Demonstrate how to glue down two or three blocks in your design; then invite one or two students to help you complete your representation. When the paper version is complete, ask students to compare the designs.

Do our two designs look the same? Have we placed each block in the right position?

Replicating pattern block designs can be challenging for many kindergartners. Consider suggesting that students first begin with a small picture or design, one that uses a limited number of pattern blocks. As students gain more experience, their work can become quite elaborate.

Students are likely to have different techniques for copying their pattern block designs. Some may build a design with blocks, then lift off one block at a time and replace it with its paper shape. Other students may use their design as a model, making the paper replica directly below or next to the original design. If you notice some students who begin to add to their block design as they glue down the paper shapes, suggest that they also add the new blocks to their original design, so the paper design and the block design are exact copies of each other.

Observing the Students

As students work with the pattern blocks, consider the following:

- How do students use the pattern blocks? Do they lay the blocks flat to make designs or pictures? Do they stand them on edge or build vertically with them? Do they stack them?
- How do they describe the different pattern blocks? Do they refer to them by shape? by color?
- Do they have a sense that pairs or combinations of blocks can be substituted for other blocks? For example, do they substitute two trapezoids for a hexagon, or two triangles for a blue rhombus?

If students are making representations of their work, consider further:

- How do students approach the task of replicating their design? Do they replace one block at a time? Do they build next to or under their original design?
- Are students able to make a copy of their design? Is this a challenging task? Can they tell when a block is not in the correct position?

Sharing Ideas While pattern blocks are an available choice, students may be interested in sharing the designs they are making. If they are also making representations on paper, they can bring these to share in group meeting. Otherwise, since their work will not be very portable, you may want to set aside time for the class to walk around and look at work in progress. The end of a session, just before clean-up, is a good time for this.

Variation

Students can count the number of total blocks or the number of each shape they used in their pattern block designs. You could set up a recording sheet with a chart of the six pattern block shapes (see below), or suggest that students find a way to record the information themselves.

Shape	⬡	⬠	◊	□	╱╱	△	Total blocks
How many?							

Choice Time

Shape Mural

What Happens

Students design and make a group shape mural, based on the illustrations in *The Shape of Things*. Their work focuses on:

- using shapes to design a picture
- relating 2-D shapes to real-world objects
- putting shapes together to make other shapes
- exploring relationships among shapes
- developing vocabulary to describe and name shapes

Materials and Preparation

- Continue to make available *The Shape of Things* or other shape book.
- Prepare the paper (butcher paper, oak tag, or other large sheet) you will be using for the mural. You might break up the space with divisions for land, water, and sky, showing an environment familiar to your students (for example, city/lake; farm land/river; beach/ocean). Alternatively, you might present the blank mural and allow students to decide how to organize the space.
- Provide a manufactured set of paper pattern blocks or prepare some from the Pattern Block Cutouts (p. 160), enlisting parent volunteers to help. Store each shape in its own container (resealable plastic bag, small box). **Note:** Students tend to use *lots* of hexagons.
- Provide the Shape Cutouts remaining from students' work on the Book of Shapes, making extra copies as needed.
- Make available coloring materials and glue sticks or paste.
- Plan to use chart paper or the board to record students' ideas for their mural during a brainstorming session.

Activity

Look again with the class at the last two facing pages in *The Shape of Things,* or at suitable pages in other shape books you have presented.

The woman who illustrated this book used lots of different shapes to make a scene with water and land and sky on these two pages. What kinds of things did she put in her picture? (Castle, ice cream truck, clown, kite, train, restaurant, houses, beach)

Can you see what shapes she used to make each thing? Alexa says she used a lot of rectangles and circles to make the train. Why do you think she used those shapes for a train?

Show students the paper that will become a mural of shape pictures. As necessary, spend a few minutes discussing what a mural is.

We're going to be making our own shape mural together. This is the paper we will use. It's mostly empty now, with just three spaces for land, water, and sky. We'll use paper shapes like the ones we used to make our Book of Shapes, and also paper pattern blocks. Let's think together about what we could make with these shapes to go on the mural.

As students generate ideas, record them in categories that match the space divisions on the mural, such as Sky, Land, and Water. This can simplify management later; it also helps students plan ways to use the whole space and to think about the overall composition.

Ask one or two students to model how they would make their suggested picture with the shapes. Ask them to tell the group which shapes they are choosing and why. These examples can be the first contributions to the mural, giving you a chance to demonstrate how students will attach their creations to the mural paper.

You can always expand the set of shapes available by cutting out additional kinds of shapes. One approach is to challenge students to make the entire mural using only paper shapes, no drawing materials. Alternatively, offer art supplies (coloring materials, paint, rubber stamps, stickers) to provide a variety of ways to design and add to the mural.

Explain that students can take turns working in small groups on the shape mural during Choice Time. Since this is a group project, students may have different or conflicting ideas about what to put on the mural. You will need to lead the group to agree that students can make whatever changes they want to their *own work,* but everyone will respect other people's work by leaving it as is.

Observing the Students

As students are working on the shape mural, ask them to tell you about the shapes they are using and what they are making.

- How do students talk about and describe the paper shapes? Do they use names? informal language? color? size? attributes of the shapes?
- How do students decide which shapes to use in a designing a particular object or scene? Do they relate the shape of the 3-D object to the 2-D shape they use? ("Tires are round, so I'm going to use circles to make the wheels on my car.") Or do they look at a 2-D shape and assign it to an object? ("This oval looks like a tree.")
- What attributes of shape (number of sides, relationship between those sides, size, orientation) do students comment upon in the process of construction? What language do they use to describe these attributes?
- Do students put shapes together to make other shapes? (For example, do they use two trapezoids to make a hexagon, or two rectangles to make a square?)

At different times while the mural is under construction, ask the group as a whole to consider it. Which shapes do they see? Which shape has been used the most? How many different shapes have been used?

Variations

- Students might like to make another mural, or add more paper to extend the existing mural.
- Students design their own individual shape collage, using the same assortment of paper shapes and a smaller base sheet of paper.
- Students design the side of a house, based on the examples pictured in the book *My Painted House, My Friendly Chicken and Me,* by Maya Angelou (Clarkson Potter, 1994).

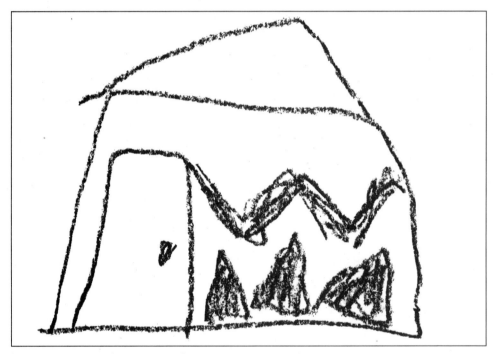

A student's design for the side of a house, using 2-D shapes.

Learning to See Shapes in the Environment

Teacher Note

Students come to the study of geometry with a great deal of practical experience, as they have been seeing and interacting with shapes in their environment all their lives. This unit helps students observe more carefully, focus on the characteristics of different shapes, and begin to describe the similarities and differences they see. The early activities in the unit start with the knowledge of shapes that students bring to the classroom, and then provide opportunities for them to expand and build on what they already know about shapes.

As students observe and describe shapes, your role is to help them consider and observe closely a wider range of shapes. For example, students should see many examples of triangles of different sizes, different shapes, and oriented in different positions:

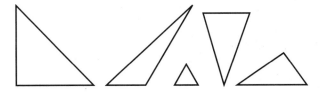

If students encounter only the triangle block in the pattern block set, they may think the term *triangle* describes only equilateral triangles (triangles with equal sides).

As they examine objects and pictures in the classroom and at home, they have the opportunity to see many kinds of triangles, and they begin to expand their understanding of what a triangle is. At this early age, all students won't necessarily agree on which shapes are triangles, but a sound and thorough understanding of what a triangle is—which will come in the later elementary grades—grows out of looking closely at, describing, and comparing many different shapes. The **Teacher Note,** How Young Children Learn about Shapes (p. 22), offers more information on this.

Through the emphasis on finding shapes in the environment, students become much more aware of the wide variety of shapes they see every day, and they begin to describe more carefully the characteristics of those shapes. Once students get into the habit of noticing shapes, they are alert and enthusiastic in telling the class about the shapes they see:

> "That tree could be a cylinder!"
>
> "We have tiles on our bathroom floor that are just like that yellow shape in the pattern blocks."
>
> "They put up big cone things on our street where they're fixing it."

Teacher Note: How Young Children Learn About Shapes

Geometry in the lower elementary grades is a lot more than learning to say the names of shapes. Young children certainly should hear words for common shapes—*square, circle, triangle, rectangle, sphere, cube, cone, pyramid,* and so forth—used correctly in context. However, building meaning for these words involves much more than seeing a few examples and memorizing the name for that shape. To give you a sense of how difficult developing a thorough meaning for shape names can be, consider these two experiences, examples of what upper-grade teachers commonly report:

- Some fifth graders have trouble identifying an obtuse triangle as a triangle. Even though they can give an accurate definition of a triangle, they don't believe that the shape below is a triangle because it doesn't look the way they think a triangle should look.

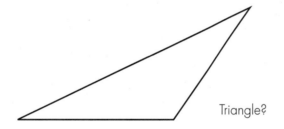

Triangle?

- Some third graders decide that a 5-by-6 array of tiles should be called a square because it "looks squarish." Again, these third graders "know" the definition of a square, but what they "know" and what they believe in their hearts about squares are two different things. They have not yet completely thought through how to connect the definition with a range of examples.

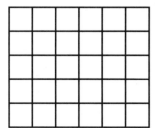

Square?

Kindergarten students are just beginning to learn which shapes are described by words such as *square, circle, rectangle, triangle, cube,* and *sphere.* To do this, they have to figure out what characteristics make a difference in the classification system we use. For example, *size* and *color* don't matter when we classify a shape as a rectangle or a circle. Students seem to understand this quite early—a *big* triangle is still a triangle, a *small* circle is still a circle. However, they may think that orientation does matter, so that a tilted square is not a square any more—now it's a diamond. A shape that has some characteristics of a circle may seem to them to be a circle, even though it is actually an oval.

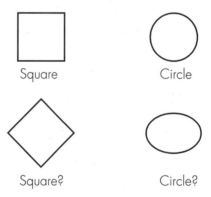

At this age, students often call 3-D shapes by the names of 2-D shapes that have some similarities; for example, they might call a sphere a *circle* or call a pyramid a *triangle.* Keep in mind that these students are noticing something important: They are beginning to pay attention to the characteristics of these shapes. There is something triangle-like about a pyramid and something circle-like about a sphere. Part of the development of geometric knowledge is moving from seeing shapes as wholes to becoming more and more competent at analyzing their characteristics and making decisions about which characteristics matter in which situations.

Teacher Note
continued

Think of a very young child learning about animals. At first, the child may call every four-footed creature "doggy." The parent may gently say, "Oh, you mean the *horse*," but for a while the child still persists in calling the horse "doggy." After enough experiences with horses—in real life, in books, or on TV—the child begins to notice the characteristics that make horses different from dogs. Yes, horses do have four legs and tails just like dogs; they might be the same colors as dogs; they have long faces, as many dogs do. But certain things about them are quite different: Their ears are always a certain shape and quite small in proportion to their heads; they are much larger than dogs, close to the height of an adult person; they have characteristic tails, and a certain way of switching them. We don't always know exactly what is being recognized when the child finally says "horse" instead of "doggy," but finally the child puts together how one set of attributes fits "dog" while a different set of attributes fits "horse."

This example suggests two things: (1) that students need a rich base of experience on which to base the development of meaning for language, including mathematical language, and (2) that interaction with adults is part of what helps them begin to sift through and organize these experiences. As you talk with students while they are working with shapes, enter into their conversations by using the same terms they are using. At the same time, help them develop their language by asking questions or making comments that challenge them to be clearer and more precise. The following interactions demonstrate how you might do this. In the first example, the teacher calls attention to the idea that even though blue and tan pattern blocks are not identical in shape, both can be described by the same shape name.

Shanique: I'm making a path with the diamonds.

Are you going to use all blue diamonds, or are you going to use some of the tan diamonds?

In a similar manner, the teacher points out to another student that the cubes ("square ones") in the Geoblock set come in several sizes.

Ravi: I'm using the square ones to build a wall.

There are lots of Geoblocks like that. Are you going to use the tiny square ones or the bigger ones?

You can also introduce conventional mathematical names for shapes, so that students hear these terms used in context. For example, here the teacher introduces the term *hexagon* for the yellow pattern block:

Felipe: When I put three blue ones on top of the yellow one, they just fit.

Felipe noticed that three of these blue blocks can fit right on top of the hexagon. Did anyone notice any other blocks that can fit right on the hexagon?

In an interaction at the Geoblock station, the teacher introduces the term *cubes*:

Tiana: I need more tiny boxes for the top of my castle.

Tiana is looking for tiny cubes. See how she's using them on her castle? Does anyone have some more of the smallest cubes that Tiana could use?

Students are not expected to use the geometric terms in kindergarten. They will begin to learn them naturally, as they learn other vocabulary—by hearing them used correctly in context. Throughout the elementary grades, students will have many experiences in classifying, describing, and defining shapes of both two and three dimensions.

Teacher Note: About Pattern Blocks

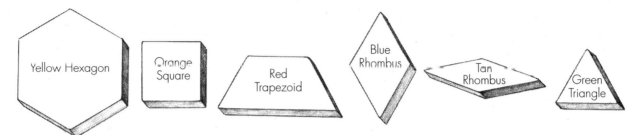

The pattern block set is made up of six geometric shapes in six colors.

- *Yellow hexagon.* A hexagon is a six-sided polygon. This block is a regular hexagon because all the sides and angles are equal.
- *Red trapezoid.* A trapezoid is a four-sided polygon that has one pair of parallel sides.
- *Green triangle.* A triangle is a three-sided polygon. This triangle is an equilateral triangle because all the sides and angles are equal.
- *Orange square.* A square is a four-sided polygon with four equal sides and angles. (A square is also a rectangle.)
- *Blue rhombus and tan rhombus.* These two shapes are both parallelograms—four-sided polygons with two pairs of parallel sides. They are a special kind of parallelogram, called a *rhombus*, that has four equal sides. (You might notice that the orange square is also a parallelogram with four equal sides, so it is actually a rhombus, too.)

Young children typically refer to the blue and tan blocks as *diamonds*. There is no need to discourage student use of this familiar term. As long as they are communicating effectively, let them use the language they are comfortable with while you continue to model use of the mathematical term *rhombus*.

The pattern blocks are related in a variety of ways. The length of every side is equal, except for the long side of the trapezoid, which is twice as long. The area of the blocks is also related. Two red trapezoids, three blue rhombuses, and six green triangles all have the same area as a yellow hexagon and can be arranged into a shape that is congruent with the yellow hexagon.

Thus, a red trapezoid is half a hexagon; a blue rhombus is one-third of a hexagon, and a green triangle is one-sixth.

Pattern blocks come in plastic bins for convenient classroom storage. If sharing is an issue, use smaller tubs or shoe boxes to split each set into smaller sets. The original set provides enough blocks for four to six students. You can limit the total number of blocks students may use by providing either a scoop for portioning out materials or a mat on which to work, thus limiting the amount of space they can cover.

Pattern blocks are used in the *Investigations* curriculum at every grade level. In kindergarten, pattern blocks are first introduced in the unit *Mathematical Thinking in Kindergarten,* and students use them to create and extend patterns in the unit *Pattern Trains and Hopscotch Paths.* In the data unit, *Counting Ourselves and Others,* students grab handfuls of these blocks and find a way to represent what they grabbed on paper. In this geometry unit, *Making Shapes and Building Blocks,* students use pattern blocks to fill in puzzle outlines. They also explore many ways to "make" the hexagon as they play a game called Fill the Hexagons. The *Shapes* software used in the geometry unit also has activities with pattern block shapes.

The pattern block set in the computer software includes the six standard pattern block shapes and a seventh shape, a quarter circle. The straight sides of the quarter circle are the same length as the sides of the square, so that it fits easily with the other shapes. The addition of the quarter circle extends students' exploration of shape to include semicircles, circles, and other curved shapes.

DIALOGUE BOX

Ideas for a Book of Shapes

As this class starts their investigation of geometry, they look at the book *The Shape of Things*. Building on what students know and notice about the shapes in the book, the teacher introduces the idea of looking for shapes in real-world objects.

This is the book we're going to look at today. What do you think it might be about?

Thomas: Snow. *[He points to the cover.]*

Thomas thinks maybe the book is about snow, because of these small white dots on the cover. Any other ideas?

Ayesha: A house, because I see a house.

What if I told you that the name of this book is *The Shape of Things*?

Justine *[gasps, and her arm shoots up]*: Shapes!

[Smiling, the teacher opens the book.] **What do you notice here?**

Kadim: There's a lot of different shapes.

Tess: I know a shape that's not there, a hopsigon.

A hexagon is another kind of shape, you're right.

Carlo: Hey, it's a pattern! *[He points to one row.]* Purple, yellow, purple, yellow . . .

I see another pattern in this same row. Does anyone see something besides a color pattern? . . . Remember the name of this book, *The Shape of Things*.

Jacob: There's a rectangle-triangle pattern.

You said that so clearly, I know exactly which one you mean. *[The teacher points to a row of alternating rectangles and triangles.]*

Shanique *[pointing to another row]*: The circle and the oval, the circle and the oval.

Yes, there's another pattern of shapes.

After this introductory discussion, the class reads the book together. Several students notice that "it's a rhyming book," while others comment on the shapes in the pictures. The teacher then talks about making a class Book of Shapes, starting with colored paper Shape Cutouts.

You're going to pick one of these shapes. You'll glue it somewhere on your paper, and then you'll make it into something, just like the illustrator in this book did.

What was a shape you saw on one of the pages in this book? . . . A circle. Yes, the illustrator took a circle and made a Ferris wheel. If you chose a circle, what else could you make, besides a Ferris wheel?

Tess: An eye.

Xing-Qi: A ball.

Charlotte: Glasses.

Luke: An apple.

Gabriela: Cherries.

Henry: I want to do a necklace with one of those. *[He points to a diamond.]*

The diamond looks like a jewel to you?

Brendan: I'm going to use a rectangle to make a stamp.

I know you're really interested in stamps, Brendan. You might even use another rectangle as an envelope.

Maddy: Or you could do a door.

Alexa: I'm going to use the egg one [oval] to make a mirror.

Kylie: I think a triangle would make a good hat, a Mexican hat.

Tarik: Or a shark.

Those are good ideas. Triangles always remind me of sandwiches. You can make whatever you want with your shape to create a page for our class Book of Shapes.

INVESTIGATION 2

Exploring Shapes with the Computer

Focus Time

Introducing the *Shapes* Software (p. 28)
Students are introduced to the basics of computer use as well as some of the special features and tools of the *Shapes* program. Rather than planning a standard whole-group Focus Time session, schedule time to work with small groups at the computer to introduce *Shapes* while others are working on Choice Time activities, including those continuing from Investigation 1.

Choice Time

Free Explore with *Shapes* on the Computer (p. 31)
Students work on the Free Explore activity, using the tools in the *Shapes* software to make designs and pictures with pattern blocks.

Pattern Block Puzzles (p. 34)
Students use pattern blocks to fill in puzzle outlines.

Continuing from Investigation 1
Book of Shapes (p. 12)
Shape Mural (p. 17)

Mathematical Emphasis

- Exploring the *Shapes* software with a partner
- Visualizing how to move a shape so that it is oriented correctly to fit into a design
- Finding combinations of shapes that fill an area
- Building knowledge about the relationships among pattern block shapes
- Developing vocabulary to describe 2-D shapes

Teacher Support

Teacher Notes
Managing the Computer Activities (p. 36)
Why Use Pattern Block Shapes on the Computer? (p. 37)

INVESTIGATION 2

What to Plan Ahead of Time

Focus Time Materials

Introducing the *Shapes* Software
- Computer(s) with *Shapes* software installed

Choice Time Materials

Free Explore with *Shapes*
- Computer(s) with *Shapes* software installed

Pattern Block Puzzles
- Pattern blocks: 1 bucket per 4–6 students
- Pattern Block Puzzles 1–10: 4–6 copies of each (duplicated on tagboard, if possible)
- Paper pattern blocks: manufactured, or make your own (optional)

Book of Shapes
- Shape Cutouts in a variety of colors, large paper, coloring materials, art supplies, and glue sticks from Investigation 1
- A stapler or hole-punch and string, for binding the class book (optional)

Shape Mural
- Mural as started in Investigation 1
- Paper pattern blocks and Shape Cutouts
- Coloring materials, art supplies, glue sticks or paste
- *The Shape of Things* by Dayle Ann Dodds (for student reference)

Preparation for Computer Use

Investigation 2 introduces the *Shapes* program (provided on the disk with this unit). The computer activities for this unit are optional but highly recommended. For more information, see the **Teacher Note**, Why Use Pattern Block Shapes on the Computer? (p. 37). If you cannot do the computer activities, skip Investigation 2 and introduce Pattern Block Puzzles as a Choice Time activity for the next investigation.

Before starting this investigation, install *Shapes* on the computers you will be using and try each of the activities. Going through the *Shapes* Teacher Tutorial (p. 117) helps you get started.

Read the **Teacher Note**, Managing the Computer Activities (p. 36). If you will be using computers located in your classroom rather than in a lab, you will need a system for cycling students through activities at the computer. Some teachers use a class list as a sign-in sheet to keep track of which students have had a turn. Students don't take a second turn until all the names on the list have been checked. Other teachers use name pins—clothespins labeled with student names—clipped onto a large can or basket. When students visit the computer, they take their name pin off and put it in the container. The pins left clipped to the edge represent students who still need to visit that activity choice. Your students may have other ideas to help set up a fair system so that everyone gets an equal chance to use the computers.

Focus Time

Introducing the *Shapes* Software

What Happens

After a brief whole-class introduction to the computer, you will work with small groups at the computer to explain the use of *Shapes*. Students' work focuses on:

- exploring the *Shapes* software on the computer with a partner
- visualizing how to move a shape so that it is oriented correctly to fit into a design

Materials and Preparation

- Have *Shapes* installed on each computer you will be using. Before teaching your students how to use *Shapes*, you should have gone through the *Shapes* Teacher Tutorial (p. 117) and tried the activities yourself on the computer.
- Set up a class list or other method for keeping track of computer use.

Activity

Introducing the Computer

The computer is a tool we can use to explore mathematics. We are going to work with a program called *Shapes* as we continue to investigate shapes, like squares and triangles. This program lets you build with pattern blocks on the computer.

Explain your expectations for pairs of students working at the computer. If computer use is new to your class, model how partners might share the computer and the mouse, taking turns as they do when they are playing a game or working together on other activities.

Some students will likely have some computer experience already and will be able to manipulate the mouse and the shapes fairly easily, while others will have little computer experience and will need help from their peers and from you. Encourage students to work and talk together about their work on the computer, and to help each other when they are stuck.

Talk with students about the system you have set up to ensure that everyone gets a chance to use the computer. For example, if you have a sign-in sheet to monitor computer use, introduce and explain it, and model its use.

I will be introducing the software to small groups during Choice Time, while everyone else is working on other choices. Don't worry if it's not your turn the first time, because everyone will get a turn to work at the computer and explore the *Shapes* software.

Depending on your daily schedule, you might begin a Choice Time now, introducing your first small group to the software while other students work on the other Choice Time activities. See the Choice Time activity, Free Explore with *Shapes,* for suggestions on introducing the software to small groups of students.

Note: The following discussion is intended for use after all students have had a chance to work on Free Explore with *Shapes*.

Activity

Sharing Work in Free Explore

Students will be working in pairs on the *Shapes* activity Free Explore throughout the rest of this investigation. During this time, hold several whole-group or small-group discussions when students can briefly share some of their work. Some might like to show printouts of designs or pictures they made on the computer. Others might like to share discoveries they made—about what particular tools do, or how particular shapes go together. Periodically, you might leave some computer creations showing on the screen for other students to walk by and view.

Focus Time Follow-Up

 **Choice Time**

Four Choices Working in pairs on the computer, students make pictures and designs with pattern blocks in the Free Explore activity (p. 31). In an off-computer activity, Pattern Block Puzzles (p. 34), students fill in outlines (puzzles) with pattern blocks. In addition, students continue working on two choices from the previous investigation, Book of Shapes (p. 12) and the Shape Mural (p. 17). As necessary, continue to offer a choice for Exploring Clay, so that students have sufficient time with free exploration before working with the clay or playdough in Investigation 4.

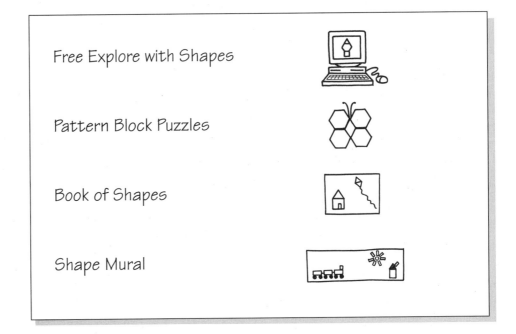

Choice Time

Free Explore with *Shapes* on the Computer

What Happens

Students become familiar with the *Shapes* software and its tools as they use pattern blocks to make pictures or designs on the computer. Their work focuses on:

- exploring the *Shapes* software with a partner
- visualizing how to move a shape so that it is oriented correctly to fit into a design

Materials and Preparation

- Be sure *Shapes* is installed on each computer you will be using.
- Have in place your class list or other method for keeping track of computer use.

Activity

Gather a small group of students around the computer with the largest monitor. Begin by explaining that this piece of computer software lets them work with pattern blocks on the computer screen.

The first activity on the computer is called Free Explore. It is a lot like what you have been doing with the pattern blocks—having a chance to play and build and design with them, and to figure out what kinds of things you can make with them, and to see what you notice about them.

Ask what kind of things students have been designing or building with the pattern blocks. You can use those ideas to demonstrate the *Shapes* tools.

Show students how to turn on the computer and open the *Shapes* program by double-clicking on the icon. Demonstrate how to choose the activity Free Explore by clicking on it once.

Point out the Shape bar with seven shapes at the left side of the screen. Ask students what they notice about the shapes in this bar. They will likely recognize the pattern block shapes, and may or may not notice the seventh, extra shape at the bottom—a quarter circle. If no one mentions this new shape, point it out.

Then demonstrate how to use the following tools, first clicking on each one in turn in the Tool bar across the top of the screen.

Arrow　　　　　Turn　　　　　　Erase One　Erase All

- Use the Arrow tool to select shapes from the Shape bar (notice that the arrow turns into a hand) and drag them into the Work window. Move several shapes into the Work window. Point out that blocks placed side-to-side will "snap" together.
- After selecting some shapes, show how to turn and rotate them using the two Turn tools.
- Show how to erase one shape at a time using the Erase One tool.
- Show how to erase all the shapes in the Work window using the Erase All tool.

Once you have introduced these tools, students should have a chance to try them immediately. Encourage pairs to work together to decide what they would like to build and to help each other figure out how to go about actually building it. Asking (and helping) students to articulate what they are doing, or what they think needs to happen to a particular block in order for it to fit, helps students develop geometric language and become more aware of the geometric motions they are using (sliding and turning).

As students are ready, you may want to introduce some other tools, such as the Flip tool, or Duplicate, which allows them to make copies of blocks that are already on the screen. This is especially useful when you have turned a block the way you want it and want another one in the same orientation. By duplicating, you avoid having to start all over with a new block.

Many students will continue to need some personal assistance with the computer. Most of the time, their questions will require only short answers or simple demonstrations. For example, a student may not know how to use the mouse. Often students who are more familiar with computers can assist those who need help. Encourage students to experiment with the tools and then to share, with each other and with you, what they've found out. It is not unusual for students to discover things about the software that the teacher doesn't know.

Observing the Students

As you observe students using Free Explore in *Shapes*, keep in mind some basic questions about learning to use the software:

- Are students able to select blocks and move them across the screen?
- Are they using the Turn tool to turn or rotate pattern blocks?
- Can students erase both an individual block and the whole screen?

You will also want to consider their ability to visualize how blocks go together:

- What kinds of patterns, designs, or pictures do students create? How do they decide which shapes to select? Do they choose them randomly? Do they begin to plan what shapes they will need for their design and select these deliberately?
- Which tools are students using and experimenting with? How comfortable are they using these tools? Do they have a sense of what will happen when they choose a particular tool?
- Once a shape has been moved onto the screen, can students figure out how to move it into position? When a shape is in the right position but the wrong orientation for what they want, can they see that it needs to be turned? Do they begin to visualize how they will have to move and turn a shape in order to add it to their design in the way they want? When they have turned a shape one turn, can they use what they see to decide if they need to keep turning it or if they should turn it in the other direction? Do they notice what size corner, or angle, will fit in a particular location?

Toward the end of a Choice Time session, give a 5-minute warning to any students working on the computer. Those who have finished a design or picture might like to save or print out their creation. If some students have not finished, help them save their design for further work during their next turn at the computer.

Remember to give students a chance to share their work, either as printouts or on the screen, as described on p. 29.

Choice Time

Pattern Block Puzzles

What Happens

Students find ways to fill puzzle outlines with pattern blocks. Their work focuses on:

- finding combinations of shapes that fill an area
- building knowledge about the relationships among pattern block shapes
- developing vocabulary to describe 2-D shapes

Materials and Preparation

- Provide pattern blocks, 1 bucket per 4–6 students.
- Duplicate Pattern Block Puzzles 1–10, preferably on tagboard, making 4–6 copies of each available.
- For the variation that involves recording solutions, make copies of the puzzles on plain paper. Provide paper pattern blocks (manufactured or made from blackline masters) and glue sticks or paste.

Activity

Introduce this activity with two or more copies of Pattern Block Puzzle 1 and a container of pattern blocks. If your students are familiar with the pattern block shapes and how they fit together, these puzzles will need only a brief introduction.

Ask a student to place one block in the outline, and then ask for other volunteers to continue filling the shape. As necessary, discuss what it means to "fill in" the puzzle outline, leaving no white space inside the outline, and placing no blocks that extend over or outside the outline.

Explain that there are many ways to fill in each puzzle. When the first puzzle is filled in, use a second copy of the same puzzle and work with the group to fill in the design again, in a different way. (For Puzzle 1, only the center can be filled in differently.)

Show students where they can find Pattern Block Puzzles 1–10 during Choice Time. Students may work alone or in pairs, depending on the available supply of pattern blocks. After they finish a puzzle, students can look for other ways to fill in the same outline, or they might solve another puzzle. Remind them to return each puzzle as they finish with it since everyone is sharing the puzzle outlines.

Observing the Students

As you observe students working on Pattern Block Puzzles, try to get a sense of how they fill in an outline.

- How flexible are students in choosing pattern blocks to fill in an outlined shape? Do they seem fluent in finding ways to fit shapes together in the interior of the shape, where it is not so obvious which blocks to use?

- How easily do students recognize which shapes will fill a particular part of the design? For example, in Puzzle 1, do they see immediately that the "rays" around the hexagon sun (or flower) can be filled with tan rhombuses (and only tan rhombuses)? Do they see that a hexagon can fill the middle section?

- As students fill in shapes, do they plan ahead? Can they see in their mind what will happen if they place blocks in certain ways? When they place a block, do they think about what shapes they will need to fill in the remaining space?

- Do you see evidence that students know how to make the same shape in different ways, such as using two trapezoids to make a hexagon? Can students use equivalents among the shapes to help them change their designs? For example, do they substitute a blue rhombus for two triangles, or three triangles for a trapezoid?

- How do students talk about the pattern block shapes? (For example, students might call the yellow hexagon "the yellow one," "the hexagon," "the big one," or "the sun.") How do they talk about the ways they manipulate the pattern blocks? (For example, you might hear terms like *turn* and *flip*: "No, you have to turn it that way to make it fit." Or, "Here, flip it over like this.") How do they describe spaces that remain in a partially-filled outline? (For example, "The last part there is going to fit a red one, I can tell. It's the same shape.")

Variations

- Students can use paper pattern blocks, glue sticks, and Pattern Block Puzzles copied on plain paper to record their work. If students have difficulty replacing blocks with cutouts to record their solution, you might offer them two copies of the puzzle sheet—one for building, one for recording—so that they need not take apart their solution to record.

- Students can count and record the total number of pattern blocks used in their puzzle. Some students might like to find a way to record how many of each type of block they used as well. The chart shown on p. 16 is one possible recording tool.

Teacher Note: Managing the Computer Activities

Once the *Shapes* software has been introduced in Investigation 2, an activity using the computer is suggested for each Choice Time throughout the remainder of this unit. How you incorporate these computer activities into your curriculum depends on the number of computers available. With your particular computer setup, it may or may not be realistic for students to use the computers regularly during your usual math time. For example, maybe you have a computer lab available only once a week. Or, maybe you have only one or two computers in your classroom and need to schedule students' computer use throughout the day. Keep in mind that adapting the activities to your particular computer situation could affect the pacing of the unit.

Regardless of the number of computers available, we suggest that students work in pairs. This not only maximizes computer resources, but also encourages students to consult, monitor, and teach each other. Generally, more than two students at one computer find it difficult to share. Each pair should spend at least 15–20 minutes at the computer for any of the suggested *Shapes* activities.

Computer Lab If you have a computer laboratory with one computer for each pair of students, all your students can do the computer activities at the same time, rather than as a Choice Time option.

Three to Six Computers If you have several computers in your classroom, your setup will work for Choice Time. After you have introduced the *Shapes* software to small groups, pairs of students can begin to cycle through the computer activities, just as they cycle through the other choices. You may need to monitor computer use more closely than the other choices, to ensure that all students get sufficient computer time. Depending on your class size, you may need to cycle pairs through the computer choice throughout the school day, instead of just during Choice Time, to give everyone a chance at the computer.

One or Two Computers If you have only one or two computers in your classroom, students will likely need to use the computers throughout the school day to ensure that every pair has the opportunity to do the computer activities.

Using *Shapes* All Year This is the only unit in the *Investigations* curriculum for kindergarten that explicitly uses the *Shapes* software. However, we recommend that students continue using it for the remainder of the school year. With more experience, they become more fluent in the mechanics of the software and can better focus on the designs they want to make and how to select and arrange shapes for those designs. As students become ready, consider introducing some of the other tools, such as the Glue tool.

Experience with the kindergarten *Shapes* software will benefit students who will be going on to the grade 1 *Investigations* curriculum, as the grade 1 geometry unit, *Quilt Squares and Block Towns,* introduces new activities in *Shapes* that build on the students' kindergarten work.

Why Use Pattern Block Shapes on the Computer?

Teacher Note

Young children need to see, touch, manipulate, and experiment. Some teachers might wonder why we include activities with pattern block shapes on the computer when students could be using the actual pattern blocks instead: Isn't it better for students to manipulate the blocks directly rather than move the shapes on a flat screen? The *Shapes* software is not a substitute for using the pattern blocks. Rather, it extends students' work with the actual blocks, leading them to think about the familiar shapes and their relationships in different ways.

Using the *Shapes* software to make designs with pattern block shapes is actually quite different from making designs with the physical blocks. While knowledge about the shapes of the pattern blocks and their relationships is still critical, students have to figure out how to move the blocks on the screen to create their design. With actual blocks, when they need to turn one to make it fit in their design, they can physically turn the block until it works, without needing to describe to themselves what they are doing. They need not be explicit about how much they need to turn the block, nor do they need to work very hard to visualize how to turn it; they can just experiment directly until it looks right. However, when they use the *Shapes* software, students have to make deliberate decisions about moving or turning or flipping a pattern block because they have to select the right tool to make the movement they want. Since it is not so easy to simply try one piece after another to see if it fits in a certain place in their design, students begin to take more time to visualize which shape they really want in any given position.

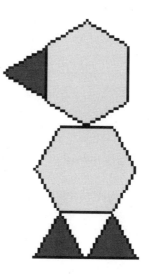

For puzzle 5 in The Shape of Things, students must rotate the hexagon one turn to fill the head. Other puzzles require even more turns.

Thus, as students use the *Shapes* software, they begin to focus on how shapes look as they are moved and turned, and what parts of the different shapes fit together. This encourages students to become more deliberate in planning a design. Instead of just selecting a block randomly and moving it around until it fits somewhere, students start to think about which block they want and how to move it into the right position. Your students will still use plenty of trial and error as they move and turn the pieces on the screen. For example, when they rotate a piece, they will probably not think about the difference between directions of rotation or the amount of each turn. They will think about the turns visually, seeing the results of each rotation as they make it and deciding on further adjustments based on what they see.

Teacher Support ■ **37**

INVESTIGATION 3

Looking at 3-D Shapes

Focus Time

3-D Shapes in the Classroom (p. 40)
Students first look at the three-dimensional shapes in a set of eight geometric solids: cube, sphere, cylinder, cone, pyramid, and three prisms (triangular, rectangular, and square). They then discuss what familiar objects in the environment each of these 3-D shapes reminds them of. Starting an ongoing Shape Hunt, students look for examples of those eight shapes around their classroom and the school.

Choice Time

Shape Hunt (p. 46)
Students continue their Shape Hunt, looking for 3-D shapes in their environment that resemble each of the geometric solids introduced during Focus Time. They use a recording sheet to keep track of the shapes they find.

Exploring Geoblocks (p. 48)
Students engage in free play with the Geoblocks, a set of three-dimensional wooden blocks.

The Shape of Things on the Computer (p. 50)
Using the *Shapes* software, students solve pattern block puzzles on the computer, exploring the ways 2-D shapes fit together to make other shapes that represent real-world objects.

Continuing from Investigation 2
Pattern Block Puzzles (p. 34)

Mathematical Emphasis

- Recognizing shapes in the environment
- Developing vocabulary to describe 2-D and 3-D shapes
- Becoming familiar with the names of 2-D and 3-D shapes
- Relating a 3-D object to a 2-D picture of its geometric shape
- Exploring 2-D shapes with the *Shapes* software
- Finding combinations of shapes that fill an area
- Building knowledge about the relationships among pattern block shapes
- Picturing the shape that will fit a particular space or design
- Visualizing how a shape needs to be moved or turned in order to fit into a particular space or design

Teacher Support

Teacher Notes
About Geoblocks (p. 55)

Dialogue Boxes
It Looks Like a Ball (p. 54)
Solving Computer Puzzles (p. 58)

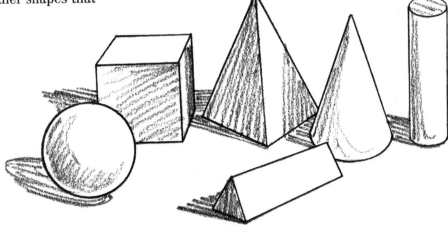

INVESTIGATION 3

What to Plan Ahead of Time

Focus Time Materials

3-D Shapes in the Classroom

- Six geometric solids—cube, cylinder, sphere, triangular prism, cone, pyramid—available as a commercial set. Alternatively, some of these shapes can be found among the wooden unit blocks in many kindergarten classrooms.
- One rectangular prism (Block J, 2 cm by 8 cm by 4 cm) and one square prism (Block K, 4 cm by 8 cm by 4 cm) from the Geoblock set

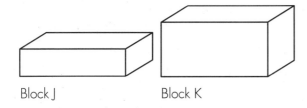

Block J   Block K

- Student Sheet 1, Shape Hunt: 1 per student
- Chart paper (optional)
- "Clipboards" made from a piece of stiff cardboard and a paper clip: 1 per student (optional)

Choice Time Materials

Shape Hunt

- Set of eight geometric solids (including two Geoblocks), as assembled for Focus Time
- Copies of Student Sheet 1, Shape Hunt, and clipboards (if used) from Focus Time

Exploring Geoblocks

- Geoblocks: 2 sets per classroom, divided into smaller sets (see the **Teacher Note**, About Geoblocks, p. 55)
- Shoe boxes or other containers for storing smaller Geoblock sets

The Shape of Things on the Computer

- Computers with *Shapes* installed

Pattern Block Puzzles

- Pattern blocks: 1 bucket per 4–6 students
- Pattern Blocks Puzzles 1–10 from Investigation 2, both tagboard and optional paper copies
- Paper pattern blocks: manufactured, or made from masters (optional)
- Glue sticks or paste (optional)

Family Connection

- Shape Hunt at Home: 1 per student for optional homework

Investigation 3: Looking at 3-D Shapes ■ **39**

Focus Time

3-D Shapes in the Classroom

What Happens

Students look at eight different geometric shapes as they discuss what familiar objects in the environment these 3-D shapes resemble. Students then go on a Shape Hunt, looking for examples of each shape in their classroom and, if possible, elsewhere around the school. Their work focuses on:

- recognizing shapes in the environment
- observing and describing 3-D shapes
- relating 3-D shapes to real-world objects
- relating a 3-D object to a 2-D picture of its geometric shape

Materials and Preparation

- Assemble a set of eight geometric solids, including a commercial set of six (cube, cylinder, sphere, triangular prism, cone, pyramid) plus a rectangular prism (Block J) and a square prism (Block K) from the Geoblock set. The **Teacher Note,** About Geoblocks (p. 55), can help you identify Blocks J and K.
- Duplicate Student Sheet 1, Shape Hunt, one copy for each student. (If you are planning to continue the hunt as homework, also duplicate Shape Hunt at Home, p. 177, for each student.)
- Have chart paper available or use the board for noting students' ideas on what the different shapes resemble.

40 ■ *Investigation 3: Looking at 3-D Shapes*

Activity

Looking at 3-D Shapes

Display the set of geometric solids where everyone can see them, and gather students for a discussion of the eight shapes.

Note: Your students are not likely to know the conventional mathematical terms for these shapes. As you introduce each shape, model the correct geometric name, but do not insist or expect kindergarten students to use the mathematical term. Students will learn the names over time, the way they learn other vocabulary, as they hear you use the terms with various examples of the shapes.

To start, choose one of the shapes for students to focus on, such as the sphere or cone.

Look very carefully at this shape. It's called a sphere. Does the shape of this block remind you of anything in our classroom or anything that you have seen outside or at home?

Students are likely to suggest such familiar objects as a ball, the globe, an orange, a scoop of ice cream, a marble, and so forth. You may want to list their ideas on chart paper and add to the list later as students find other objects during their Shape Hunt. As students suggest each similar object, ask them to tell how the shapes are the same.

Lots of people have said this shape looks like a ball. What is it about this shape that looks like a ball? . . . Felipe says that it is round. Tiana says that it has curves and is smooth all around. And Kylie says that it can roll. All of these are important things about balls and spheres.

When young children describe shapes, it may be difficult for them to name specific attributes of the shape; instead, they will talk about what it looks like or give its function. For the sphere, some students will say "It is round," while many others will probably say things like "It's a ball," or "It looks like somebody's head." Encouraging students to think in these ways about the characteristics of a shape helps them build an understanding of what makes one shape different from another. For more information, see the **Teacher Note,** How Young Children Learn About Shapes (p. 22).

After students have generated several ideas about the first shape, turn their attention to a few of the other shapes, one by one.

Luke says that this cone *[hold up the cone]* reminds him of a hat. Who has a different idea about what this shape looks like? . . . Ida thinks it looks like a pine tree. What else? . . .

Focus Time: 3-D Shapes in the Classroom

OK, let's think about a different shape. Who has an idea for a shape that we haven't talked about yet?

The **Dialogue Box,** It Looks Like a Ball (p. 54), is an excerpt from this discussion in one kindergarten class. We recommend discussing no more than two or three shapes at this time; students might describe and discuss the remaining shapes at various points during your Choice Time sessions.

Activity

Shape Hunt

Hold up a copy of Student Sheet 1, Shape Hunt, and have at hand the eight geometric solids. As you point to each shape pictured on the student sheet, ask for a volunteer to find the 3-D solid block that the picture represents.

When you look around our classroom, you might notice that things in the room have lots of different shapes. Some of them look a lot like the shapes of these blocks. You're going on a Shape Hunt, and you'll be looking for things in our classroom that have shapes a lot like the different shapes on this paper.

Use this paper to keep track of what you find. When you find something that looks like one of these shapes, put a mark next to that shape. Later, if you find something else with the same shape, make another mark.

Using a couple of examples, demonstrate for students how to record the shapes they find on the student sheet. For example, they might use X's, tally marks, or check marks. Some students may want to try writing the name or drawing a picture of objects they find, while others might also use numbers to count how many.

I think the aquarium in our meeting area looks like one of these shapes. Which shape do you think it looks like? Where should I put an X? . . . What about our pencil can? Does it remind you of one of these shapes?

Distribute Student Sheet 1, Shape Hunt, to each student, along with the clipboards you have prepared (if any). If you have a parent volunteer, you may want to extend the Shape Hunt outside of the classroom, into other parts of the school or an outside play yard. In any case, point out that some of the shapes may be very hard to find. Although students may not find all eight shapes pictured on the student sheet, they should find as many different shapes as they can.

The Shape Hunt can be done as a whole-group activity for the rest of Focus Time, or it could be done by smaller groups as part of Choice Time while other students continue working on the choices in Investigation 2. In either case, it remains a Choice Time activity for the rest of Investigation 3.

Observing the Students

Observe and interact with students while they are on the Shape Hunt.

- How do students talk about and describe shapes? Do they use a 2-D name to refer to a 3-D shape? For example, do they say "circle" for a sphere or "square" for a cube?
- Are students able to match the 3-D block to its 2-D picture?
- Are students able to look at an object and relate it to a similar shape on their student sheet? If they are having difficulty, suggest that they choose one block to carry with them for reference as they hunt for that shape.
- Are students able to describe the shape of an object? Can they tell in what ways it is the same as one of the geometric solids?

Activity

Comparing 3-D Shapes

A few days after introducing the Shape Hunt, when most students have had a chance to do the activity, plan a short sharing meeting. Ask students to tell about some of the objects they have found. Before this discussion, gather a few interesting items from around the room that are representative of the 3-D shapes students have been looking for. Students can also help in collecting a few examples from around the classroom. During this meeting, you may want to add to the class list of objects resembling each geometric solid.

Focus Time Follow-Up

Looking for Shapes at Home Students can go on a Shape Hunt at home. Each student will need a copy of Shape Hunt at Home, with notes for families and a place to keep track of the shapes they find. Encourage them to remember some examples to share with the class. Students might ask their parents to help them write the words, or they might like to draw pictures of the objects they find.

Comparing Geometric Solids To help students concentrate on the attributes of individual shapes, choose two of the 3-D solid blocks for students to compare and consider more closely. For example, students might compare one or more of the following pairs. How are the two blocks similar? How are they different?

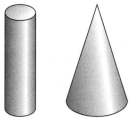

cylinder and cone

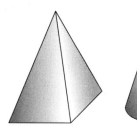

pyramid and cone

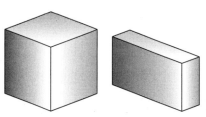

cube and rectangular prism

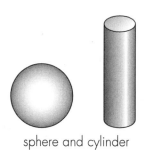

sphere and cylinder

 **Choice Time**

Four Choices During Choice Time, students continue with their Shape Hunt (p. 46). They also have time to explore the Geoblocks (p. 48). In the *Shapes* software, students work on the activity The Shape of Things, a computer version of Pattern Block Puzzles (p. 50). Students may also continue their work with Pattern Block Puzzles off the computer (the Investigation 2 activity). If computers are available in a lab but not during Choice Time, consider presenting the computer activity The Shape of Things to the whole group during your computer lab time.

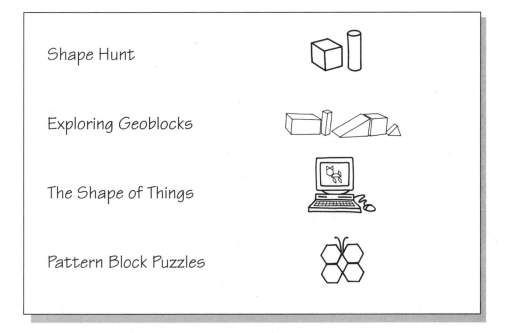

Note: If you are not using computers with this unit and have skipped Investigation 2, introduce the Pattern Block Puzzles (p. 34) as a new activity for this Choice Time. In addition, students may continue working on the class Book of Shapes and the Shape Mural, as introduced in Investigation 1. If you have been offering Exploring Clay as a choice for free exploration, continue to include it as an option here.

Focus Time: 3-D Shapes in the Classroom ■ **45**

Choice Time

Shape Hunt

What Happens

Students continue their 3-D shape hunt, looking for objects whose shapes resemble the geometric solids pictured on the Shape Hunt student sheet. Their work focuses on:

- recognizing shapes in the environment
- observing and describing 3-D shapes
- relating 3-D shapes to real-world objects
- relating a real-world 3-D object to a 2-D picture of its geometric shape

Materials and Preparation

- Students continue to use their copy of Student Sheet 1, Shape Hunt, from Focus Time, and any clipboards you may have provided.
- Continue to have available for visual reference the set of eight geometric solids you assembled for Focus Time.

Activity

Students continue to look for shapes in their classroom. With an extra adult to supervise, a small group can extend the hunt to areas of the school outside the classroom.

Briefly review any work on the Shape Hunt that was done during Focus Time. As you look together at the set of geometric solids, ask students to share some of the objects they have already found that resemble the shapes of these blocks. If students have done the optional homework with the sheet Shape Hunt at Home, they can also refer to this work.

You might ask a few students to share their methods for keeping track of the things they find. Look especially for different methods of recording, such as tally marks, pictures, words, or numbers.

Observing the Students

Observe and converse with students as they continue the Shape Hunt.

- How do students talk about and describe shapes? Do they use a 2-D name to refer to a 3-D shape? For example, do they say "circle" for a sphere or "square" for a cube?
- Are students able to look at an object and relate it to a similar shape on their student sheet?
- Are students able to describe the shape of an object? Can they tell in what ways it is the same as one of the geometric solids?

Sharing Our Discoveries Every few days, hold brief meetings when students can share their findings. You might help the class keep track of the objects they have found by recording their findings on chart paper. Some shapes, in particular the triangular prism and pyramid, may be difficult to find in real-world objects.

cone — party hat, ice cream cone, orange "road" cone, teepee, party horn	cylinder — paper towel roll, can, pencil, jumprope handle, chalk
sphere — playground ball, globe, marble, bead, moon	triangular prism — house roof, tent
rectangular prism — book, eraser, door	cube/box — aquarium, toy bin, bookshelf, shoe box
cube — game cubes, cubby, tissue box	pyramid — roof, hat

Choice Time

Exploring Geoblocks

What Happens

Students spend time in free exploration with the Geoblocks, a set of three-dimensional wooden blocks that are related to one another in particular ways. Students' work focuses on:

- exploring Geoblocks and their attributes
- using informal language to describe geometric shapes

Materials and Preparation

- Provide 2 sets of Geoblocks, separated into smaller sets. See the **Teacher Note,** About Geoblocks (p. 55), for ideas on how to sort the blocks and for more information about the shapes included in each set.

Activity

If you have been teaching the year-long *Investigations* curriculum for kindergarten, Geoblocks were introduced in the first unit, *Mathematical Thinking in Kindergarten.* Students who are familiar with these blocks and have been using them throughout the school year will need less free exploration time. In this case, after a brief exploration period, you might introduce the idea of building a wall, one of the variations to Build a Block described on p. 74.

If students have had only limited experience with the Geoblocks, we strongly recommend that they have repeated opportunities to use them in free exploration.

If Geoblocks are a new material in your classroom, you may introduce them by gathering students together in a circle and displaying a set.

These are a special type of block, called Geoblocks. Geoblocks are a tool we can use to explore mathematics. Look carefully at the blocks. What are some things you notice about them?

Until students have had opportunities to work closely with the blocks, they may remark only on obvious attributes such as color, material, and possibly shape, especially by relating them to the shapes of familiar objects in the real world. They are likely to name things they could build with the blocks. Students may also relate Geoblocks to other building blocks they have used.

Explain where and how the Geoblocks can be used and cared for, as discussed in the **Teacher Note,** Materials as Tools for Learning (p. 99).

During Choice Time, one of your choices will be to explore these Geoblocks. There are only two sets of the blocks, so you will have to take turns.

Observing the Students

Consider the following as you watch students exploring the Geoblocks.

- How do students use the Geoblocks? What sorts of constructions do they make? Do they build primarily vertically, stacking blocks? Do they build more horizontally, covering space? Do they use them for dramatic play? (For example, one group thought of the tiny cubes as "gold," which they stored in a box made of prisms.)
- How do students describe the different Geoblocks? Do they describe differences in size? Do they notice that some of the faces (they will probably call them "sides") are squares, some are rectangles, and some are triangles? What words do they use to talk about these shapes?
- Do they have a sense that pairs or combinations of blocks can be substituted for other blocks? For example, do they put two (or more) Geoblocks together to match another Geoblock?

Some students will look hard for a particular block, perhaps a duplicate of one they already have. The way the set is designed, several blocks have only a few copies. When students are having trouble finding a particular block, encourage them to think about other possible ways to make a block that is the same size and shape. This can also be a helpful solution for resolving difficulties with sharing.

Sharing Ideas While Geoblocks are an available choice, periodically gather students to talk about them. For example:

- Share specific ways that students used the Geoblocks.
- Discuss any management issues that are arising, such as sharing blocks fairly.
- Compare Geoblocks with pattern blocks: How are they the same? How are they different?
- Discuss the different types of shapes in the Geoblock set. As a whole group, they could sort the blocks into "like" groups and then describe each group. (In Investigation 4, students focus on specific attributes of the Geoblock shapes.)

Choice Time

The Shape of Things on the Computer

What Happens

Students continue to think about representing real-world objects with shapes as they use the *Shapes* software to solve pattern block puzzles on the computer. They learn about and use a variety of tools to fill puzzle outlines with pattern blocks. Their work focuses on:

- finding combinations of shapes that fill an area
- visualizing what shape to select to fill in a design
- visualizing how to move a shape so that it is oriented correctly to fit into a design
- putting parts together to form a whole
- learning to use the *Shapes* software

Materials and Preparation

- Make whatever plans or arrangements are necessary for using the computers with *Shapes* installed.
- Be sure you have worked on the computer with The Shape of Things activity before introducing it to your students.

Activity

We recommend introducing this choice to small groups at the computer, while the rest of the class participates in other Choice Time activities. However, you could also begin by briefly describing the new computer choice to the whole class.

Today I'm going to introduce a new computer activity, called the Shape of Things. It uses the *Shapes* software, which all of you have explored a little bit. This activity is a lot like the Pattern Block Puzzles you have been working on, except this time the puzzle outline will be on the computer.

Reassure students that everyone will get a chance to cycle through this activity and solve the pattern block puzzles on the computer.

At the computer, show students how to get started.

1. Open the *Shapes* software by double-clicking on the icon, and click in the center window to continue.
2. Open the Shape of Things activity by clicking on the second box, showing the filled-in puzzle outline of an animal.
3. Read aloud the directions to students, and click OK to start the activity.

Demonstrate how to drag shapes into the puzzle outline, reminding students as necessary about the Arrow, Turn, Erase One, and Erase All tools that they used during the Free Explore activity in Investigation 2.

Finally, demonstrate how to choose a new puzzle outline: Click on **Number** in the menu bar, and drag the cursor down to any number. Explain that every time Shape of Things opens, the software presents outline Number 1. Once they have completed this first puzzle, they always need to use the Number menu to select another outline.

Each time students choose a new puzzle outline, the computer will ask if they want to save the work they've done on the puzzle currently showing. Decide ahead of time if you'd like students to save their work to a disk or not, and explain this process to them now. See the *Shapes* Teacher Tutorial (p. 117) for information about saving work.

As students work on the Shape of Things activity, talk with them and encourage them to talk to each other about the shapes they are choosing and the reasons for choosing them. Verbalize the way they are moving the blocks and using the *Shapes* tools. This will help them become more aware of the geometric motions of sliding, turning, and possibly flipping. Just as important, it will help them become familiar with seeing shapes in different orientations and realizing that changing the orientation does not affect the shape's name or attributes.

Many students will continue to need some personal assistance with the computer. Usually their questions will require only short answers or brief demonstrations. Often students who are more familiar with computers can assist those who need help. Encourage students to experiment with the tools and then to share with one another what they've found out.

As students are ready, you might introduce some other tools, such as Flip and Duplicate. Duplicate, which allows students to make copies of blocks that are already on the screen, is especially useful when a block is turned just the right way and they want another identical block in the same orientation. By duplicating, they avoid having to go through the turning process all over again with a new block.

Observing the Students

Consider the following as you watch students using the *Shapes* software.

- Can students select blocks and move them across the screen?
- Are they using the Turn tool to turn or rotate pattern blocks? Are they experimenting with other tools, such as Flip and Duplicate? Which ones? How are they using them?
- Can students erase an individual block? the whole screen?

Also consider students' strategies for choosing and placing blocks and their ability to visualize how the blocks go together.

- How do students decide which shapes to select? Do they select randomly? Do they plan which shapes they need and select these deliberately? Is the shape they need obvious to them for certain parts of puzzles, such as the square "feet" or the triangle "arms" in puzzle 1?
- Do students realize there are many ways to solve a puzzle? (For additional challenge, students could look for a different way to fill the same outline.)
- Are students demonstrating any knowledge of the equivalence between the pattern block shapes? For example, do they know they can fill a hexagon outline with one hexagon, two trapezoids, three blue diamonds, six triangles, or with some combination of these?
- Having moved a shape onto the screen, can students then move it into position? When a shape is in the right position but the wrong orientation, can they see that it needs to be turned? Do they begin to visualize how they will have to move and turn a shape in order to get what they want? After turning a shape once, can students use what they see to decide if they need to keep turning it or to turn it in the other direction?

Students commonly identify the outlines in the ten puzzles as (1) a clown; (2) a barn with a silo or a house with a chimney; (3) an airplane; (4) a rabbit; (5) a baby chick or duck; (6) a person dancing; (7) a scarecrow; (8) a screw or an umbrella; (9) a tractor driving to a shed, or a train to a station; (10) a puppy or other four-legged animal in the grass. However, the way students fill a puzzle may change their view of what it represents.

Some students will use their knowledge of pattern block equivalencies to color a picture that is more interesting to them. For example, one student filled the middle hexagon in puzzle 1 with two red trapezoids, so it would look like "It's wearing a red shirt."

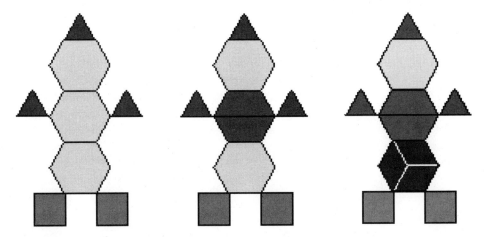

Three ways of completing puzzle 1 in The Shape of Things: (**a**) a clown in a yellow suit, (**b**) a clown in a red shirt, (**c**) a clown in a red shirt and blue pants.

For a look at two pairs of students working on and talking about a pattern block puzzle on the computer, see the **Dialogue Box**, Solving Computer Puzzles (p. 58).

Students who complete a puzzle may want to print their work. If students don't have time to complete a puzzle in one session, show them how to save their pictures on disk. Students who subsequently revisit this choice may need help finding and opening their partially completed work.

Variations

- You and the students can make your own Pattern Block Puzzles on the computer for students to solve. Open the *Shapes* software and select the activity Create a Puzzle. When you have a design, choose Save Your Work As... under the File menu, and name the puzzle (for example, "Teacher Puzzle 1" or "Robot Puzzle"). You will not see the actual puzzle outline until the next step.

  To open and then solve the puzzle, open the *Shapes* software (or choose Change Activity under the File menu) and select the activity Solve a Puzzle. In the box that then appears, choose the name of a puzzle, and an outline will appear in the window.

- You can print out designs made on the computer for students to solve with actual pattern blocks. Design and create the puzzle outline in Create a Puzzle. Then open the puzzle outline with Solve a Puzzle. To print this outline, choose Page Setup from the File Menu. In the Reduce or Enlarge box, type 180 for the percent. Choose the Color/Grayscale option. If you have a Print Preview feature, use it to check the placement of the puzzle on your page. Then print.

  Note: Different computers and printers may distort the size somewhat. Try printing first at 180%; then test your printout using actual pattern blocks. If the outline is too large or too small, adjust the percentage, experimenting until you find the size that works on your system.

DIALOGUE BOX

It Looks Like a Ball

In this kindergarten classroom, the set of geometric solids is being introduced for the first time.

Take a minute and look very carefully at this shape. It's called a sphere. Does the shape of this block remind you of anything in our classroom, or anything that you have seen outside or at home?

Tarik: It looks like a ball.

Oscar: Yeah, or a big huge marble.

Gabriela: It's a circle.

Ayesha: It looks kind of like the world we have in our room. *[She points out the globe on the bookshelf.]*

Tess: Or a planet. Planets are circles like that.

Jacob: It's kind of like a head, but heads aren't so round like that. They're kinda bumpy.

All of the things you have mentioned—a ball, a marble, a globe, a head, and the planets—do look round like spheres.

Let's try one more shape before we go on our Shape Hunt. This is a cylinder. Look very closely. Does the shape of this cylinder remind you of anything?

Carlo: Yeah! The thing paper towels come on!

Ayesha: It reminds me of the thing we use at home to make cutout cookies, if you hold it this way *[she holds the cylinder horizontally and rolls it, demonstrating a rolling pin].*

Tess: It kind of looks like a skinny can, like of food or something.

Jacob: I think it looks like one of those things that's in the front of some buildings, you know, to hold them up?

I think Jacob is thinking of the columns that some buildings have, like the front of the library when we took our class trip there. *[Jacob nods.]*

During this discussion, the teacher notices that students are beginning to pay attention to important characteristics of shapes. Although many students are not making distinctions between 2-D and 3-D shapes—a *sphere* is called a *circle*—discussions like these encourage students to carefully observe, describe, and compare the shapes they see in their environment.

About Geoblocks

Teacher Note

Geoblocks are a special set of three-dimensional wooden blocks. While similar to other kindergarten blocks (wooden unit blocks), they are smaller and are designed so that the blocks are related by volume. We live in a three-dimensional world, yet most of the geometry students do in school is concerned with two-dimensional shapes. We frequently use two-dimensional drawings to help us picture and represent three-dimensional things. For example, a blueprint provides instructions for building a house; a paper pattern, for cutting out and sewing a shirt; a diagram, for assembling a bike. One reason we include Geoblocks in the kindergarten materials is that it's important for students to work with three-dimensional materials and see the relationship between three- and two-dimensional shapes.

In the *Investigations* curriculum, students use Geoblocks in kindergarten and in the first and second grades. After the blocks are introduced in *Mathematical Thinking in Kindergarten,* they are formally used only in this geometry unit, *Making Shapes and Building Blocks.* However, we recommend that you make the Geoblocks available throughout the year, perhaps in a building area or during times such as indoor recess or free Choice Time. Most kindergartners love to build with these blocks; they build towers, towns, roads, ramps, bridges, and many other things. During this informal building time, they intuitively learn many of the characteristics of the blocks.

Because there are a limited number of Geoblocks, especially some types, sharing is often an issue. While there are lots of the smallest cubes (128), there are only two each of the largest blocks, and only four or six of many of the triangular blocks. To help with sharing, begin by separating each set of 330 Geoblocks into two equal sets, each set being enough for four or five students. Probably the easiest way to divide the main set of blocks is to find two identical blocks and put one in each set. Parent volunteers, aides, or older student volunteers can do this, although the sorting task is a good way for you to become familiar with the shapes yourself.

Encourage students to use only the blocks in their particular set. Before sending students off to explore and build, you might model ways of sharing and ways to request blocks (and to respond to requests). Students need to be able to accept the fact that the block they want may be important in someone else's construction, too. In some situations you might encourage students to combine buildings or to build together. Finally, students who cannot find a block they want can be encouraged or challenged to find two (or more) other blocks that combine to make the original. This is a possibility for most blocks in this set, which is one of their useful mathematical attributes.

Continued on next page

Teacher Support ■ 55

Teacher Note *continued*

For your own information, there are five kinds of shapes in the Geoblock set. Before reading these descriptions, we suggest that you do your own sorting of the blocks, since they are probably less familiar to you than some of the other kindergarten materials. Once you have sorted the blocks into groups that you think "go together," the descriptions will probably make more sense. Kindergarten students are not expected to learn the formal names or descriptions of Geoblock shapes.

All the shapes are *polyhedra,* three-dimensional solid shapes with flat faces. These are the five general categories:

Rectangular prisms. Prisms have two opposite faces that are the same size and shape (congruent). All other faces, connecting these two opposite faces, are rectangles. In *rectangular prisms,* the two opposite faces are rectangles, so all six faces are rectangles. Most boxes are rectangular prisms. You can also call these shapes *rectangular solids.*

Rectangular prisms

Square prisms. These are a special kind of rectangular prism. They have two opposite faces that are congruent squares. The other four faces are rectangles.

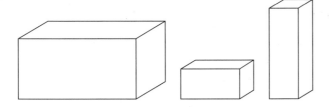

Square prisms

Cubes. Just as the square is a special kind of rectangle, the cube is a special kind of rectangular prism in which all the faces are squares.

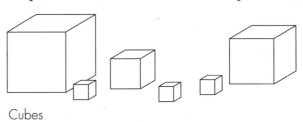

Cubes

Triangular prisms. These prisms have two opposite faces that are congruent triangles. As in any prism, the faces that connect this pair are all rectangles.

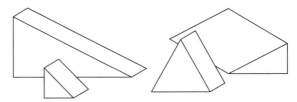

Triangular prisms

Pyramid. There is one kind of pyramid in the Geoblock set. Pyramids look different from prisms. They have one base, which can be any polygon. The rest of the faces are triangles that meet in a single point (vertex). The pyramid in the Geoblock collection is a square pyramid. It has a square base and four triangular faces.

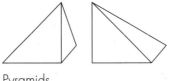

Pyramids

See the following page for a chart showing the 25 different Geoblock shapes, their dimensions, and the quantity of each shape found in a set.

56 ■ *Investigation 3: Looking at 3-D Shapes*

Cube A
1 cm × 1 cm × 1 cm
Quantity: 128

Cube B
2 cm × 2 cm × 2 cm
Quantity: 32

Cube C
3 cm × 3 cm × 3 cm
Quantity: 12

Cube D
4 cm × 4 cm × 4 cm
Quantity: 8

Rectangular prism E
2 cm × 4 cm × 2 cm
Quantity: 8

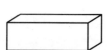
Rectangular prism F
2 cm × 6 cm × 2 cm
Quantity: 4

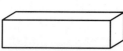
Rectangular prism G
2 cm × 8 cm × 2 cm
Quantity: 4

Rectangular prism H
2 cm × 4 cm × 4 cm
Quantity: 12

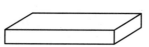
Rectangular prism I
1 cm × 8 cm × 4 cm
Quantity: 8

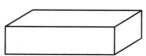
Rectangular prism J
2 cm × 8 cm × 4 cm
Quantity: 4

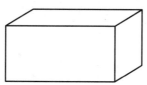
Rectangular prism K
4 cm × 8 cm × 4 cm
Quantity: 2

Triangular prism L
2 cm × 2 cm × 2 cm
Quantity: 32

Triangular prism M
4 cm × 4 cm × 4 cm
Quantity: 6

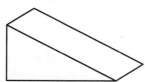
Triangular prism N
4 cm × 8 cm × 4 cm
Quantity: 2

Triangular prism O
4 cm × 4 cm × 2 cm
Quantity: 12

Triangular prism P
3 cm × 3 cm × 4 cm
Quantity: 8

Triangular prism Q
2 cm × 4 cm × 4 cm
Quantity: 2

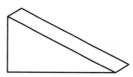
Triangular prism R
4 cm × 8 cm × 2 cm
Quantity: 6

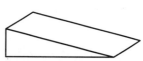
Triangular prism S
2 cm × 8 cm × 4 cm
Quantity: 6

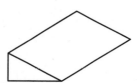
Triangular prism T
2 cm × 4 cm × 8 cm
Quantity: 6

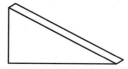
Triangular prism U
4 cm × 8 cm × 1 cm
Quantity: 6

Triangular prism V
2 cm × 4 cm × 2 cm
Quantity: 4

Triangular prism W
4 cm × 3.5 cm × 2 cm
Quantity: 6

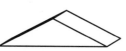
Triangular prism X
2 cm × 8 cm × 2 cm
Quantity: 6

Pyramid Y
4 cm × 4 cm × 4 cm
Quantity: 6

DIALOGUE BOX

Solving Computer Puzzles

Students in this class have just begun to work on the activity The Shape of Things in the *Shapes* software. Their conversations demonstrate two important discoveries students often make while trying to fill the puzzle outlines: (1) that shapes can be combined to make other shapes, and (2) that the orientation of a shape can be changed by rotating its position in space.

Shanique and Brendan have opened puzzle 3.

Brendan: It's a plane! *[He points to the plane's rectangular body.]* Do we have this?

Shanique: We need to get a rectangle, but there isn't any. *[She looks closely at the shape bar.]*

Together Brendan and Shanique fill in the rest of the outline, until only the plane's body is left.

Shanique: But we need to get a rectangle.

While clicking on the screen, Shanique notices that you can make a rectangle by clicking and holding down the mouse button while dragging the mouse. She thinks this may be the answer and tries to create a rectangle the right size, growing and shrinking it. But as soon as she lets go of the mouse button, the image disappears.

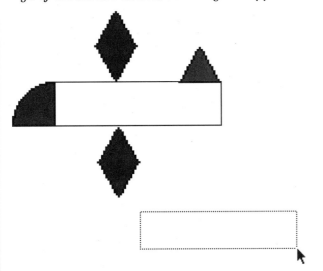

Shanique *[frustrated]*: I need to move it [the gray rectangle] up to fill the space. Should I click one of these? *[She points to the tool bar.]*

Brendan: Let me try something. *[He puts two trapezoids in the plane's body.]* Did I finish it?

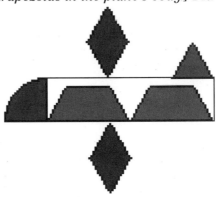

What do you think? Does it look finished?

Shanique: No. It doesn't fit right, there's still here *[pointing to the white space]*. These red ones don't work. *[She takes them out.]*

Do you think there's a way to use these shapes *[points to the Shape bar]* to make a rectangle? Remember, we talked about times when you put shapes together, and they make a new shape? Are there any other shapes, besides the red trapezoid, that you know *won't* work?

Shanique: Well, this one [quarter circle] won't, because it's curvy and rectangles are all straight.

Is there another shape you could try, one that's *straight,* more like rectangles?

Shanique: I have an idea. *[Puts a square into the end of the rectangular area.]* There, I did it!

What made you think that a square would fit there?

Shanique: Well, its goes like this *[sketching a right angle in the air with her finger]*, like an L, and that's what I needed to fit there.

At first these students could not understand how they could fill a rectangular region without a having rectangle shape of the same size as that region. As they thought about what shapes might (or might not) fit, they were thinking about the characteristics of particular shapes, and discovered that four squares could be combined to form a long rectangle.

In this second conversation, Renata and Maddy are working on puzzle 4. This is the first of the ten puzzles in which students must turn some blocks in order to orient them correctly.

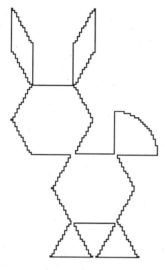

Renata: Maybe it's a rabbit.

Both girls seem clear that they need triangles for the feet and hexagons for the body. As they work, Renata calls the hexagons "stop signs" while Maddy calls them "pantagons" [pentagons]. The teacher notices this and decides to introduce the word hexagon.

I noticed that you used two yellow hexagons [pointing to the shapes on the screen] **to make the rabbit's body and two triangles to make its feet. What will you do next?**

Maddy: [choosing a tan rhombus] This is the only shape that's skinny like the ears. [She places it in the left ear of the outline, and goes back for another. After trying to place a rhombus in the right ear of the outline, she realizes it won't fit.] Hmm. I guess we're gonna hafta do that turn-y thing.

She chooses the left turn tool, and turns it once. Renata suggests checking the shape in the ear again, to see if it will fit. When it doesn't, Maddy turns the shape again, and realizes without trying that it still won't fit. After that, she tries to place the shape in the outline after each successive turn, finally resulting in a second ear.

Maddy: That was hard.

Renata: Yeah. Now all we have left is the tail. It can't be a square . . .

Renata tries several shapes, including some suggested by her partner—none of which is the quarter circle—and gets frustrated.

Are there any shapes you haven't tried yet? Is there anything special about the shape of that tail that might help you choose a shape? How did you know it couldn't be a square?

Renata: Because it goes like this [tracing the curved part of the tail] and squares don't do that.

So you noticed the curved part of the tail. Are there any shapes you could choose that have curved parts?

Maddy: Maybe this one . . . yeah, look. It's curved, too.

Renata drags over a quarter circle to put on the tail. As she realizes it won't fit the way it is oriented, she starts turning the mouse, her head, and her body, as if to will the shape to move.

Renata: It's no use.

Do you remember how you moved this shape around [pointing to the rhombus in the right ear] **to make it fit the way you wanted it to?**

Renata: Oh yeah, the circle arrow thing. [She chooses the right turn tool and turns the shape, checking it after each turn. After three turns, she finds the perfect fit.] I did it, and look, it was a rabbit. That one was hard.

These students were struggling with an important idea. As they worked to correctly orient the shapes, they were thinking not only about the characteristics of shapes (it can't be a square because it's curved), but also about the geometric motion of turning or rotating a shape to fit it into a design in a particular way.

Teacher Support ■ **59**

INVESTIGATION 4

Making Shapes and Building Blocks

Focus Time

Clay Shapes (p. 62)

Students discuss the attributes of familiar two-dimensional shapes such as a square, a triangle, a circle, and a rectangle. Then they work individually or with a partner to make outlines of these shapes with ropes of clay or playdough.

Choice Time

Clay Shapes (p. 68)

Students continue forming outlines of shapes with clay, making not only different shapes but also different examples of the same shape, such as many different triangles or rectangles.

Fill the Hexagons (p. 70)

Players roll special pattern-block game cubes and use the indicated blocks to fill in six hexagon shapes on a gameboard.

Build a Block (p. 73)

Students build one of the larger Geoblock shapes by combining two or more smaller Geoblocks.

Quick Images on the Computer (p. 75)

Quick Images is a computer activity that you introduce in an off-computer version. Students briefly view an image of two pattern blocks in a particular arrangement, then, with the original image hidden, put together pattern blocks to make a copy of the image they saw. In the computer version, students work with shapes on the screen in place of actual pattern blocks.

Continuing from Investigation 3
Pattern Block Puzzles (p. 34)

Mathematical Emphasis

- Describing and becoming familiar with the attributes of 2-D shapes
- Constructing 2-D shapes
- Finding combinations of shapes that fill an area
- Building knowledge about the relationships among pattern block shapes
- Combining smaller 3-D shapes to make a larger 3-D shape
- Analyzing visual images
- Describing position of and spatial relationships among objects

Teacher Support

Teacher Notes
Making Shapes (p. 79)

Dialogue Boxes
Three Pointy Corners (p. 80)
Circles and Ovals (p. 81)

INVESTIGATION 4

What to Plan Ahead of Time

Focus Time Materials

Clay Shapes

- Clay or playdough: one approximately 3-inch ball per student
- Sturdy cardboard mats, about 8 inches square: 1 per student
- Four shape posters, made by gluing colored copies of Shape Cutouts A–D (pp. 157–158) on large paper

Choice Time Materials

Clay Shapes

- Clay or playdough and cardboard mats from Focus Time
- Make-a-Shape Cards (pp. 178–180): 5–6 sets for the class

Fill the Hexagons

- Student Sheet 2, Fill the Hexagons Gameboard (p. 181): 1 per player or pair, duplicated on card stock and laminated (if possible), plus 1 per student, copied on plain paper, for the variation that involves recording (optional)
- Pattern blocks: 1 bucket per 4–6 students
- Pattern-block game cubes (made with blank one-inch cubes and stick-on labels, see p. 70): 2 cubes for each group of players
- Paper pattern blocks and glue sticks (optional, for recording variation)

Build a Block

- Geoblocks: 2 sets per classroom, each set divided equally in half, and with the largest blocks (square prism, 8 by 4 by 4 cm) removed as models

Quick Images on the Computer

- Computers with *Shapes* software installed
- Set of three pattern blocks (hexagon, square, triangle) in a small cup: 1 set per pair
- Small tray or piece of cardboard, and paper or cloth large enough to cover it completely

Pattern Block Puzzles

- Pattern blocks (1 bucket per 4–6 students)
- Pattern Block Puzzles 1–10 (from Investigations 2–3)
- Paper pattern blocks (optional)
- Glue sticks or paste (optional)

Focus Time

Clay Shapes

What Happens

As a whole class, students discuss the attributes of familiar shapes such as a square, triangle, a circle, and a rectangle. Then, using ropes of clay (or playdough), they work individually or with a partner to make outlines of these shapes. Their work focuses on:

- describing and becoming familiar with the attributes of two-dimensional shapes
- constructing two-dimensional shapes

Materials and Preparation

- Divide clay or playdough into balls roughly 3 inches in diameter, one ball for each student. (As needed, see p. 11 for a playdough recipe.)
- Cut squares of cardboard, at least 8 inches on a side, to make a working mat for each student.
- To make a set of four shape posters (square, triangle, circle, rectangle), copy Shape Cutouts A–D (pp. 157–158) on colored paper, cut out the shapes, and glue each shape to a separate sheet of paper. Leave enough space on the poster to add student comments.

Activity

Describing Shapes

Refer to the class Book of Shapes and the students' Shape Mural (from Investigation 1) or to the Pattern Block Puzzles to remind students of the work they have been doing with shapes.

During the last few weeks, you have been using shapes to make pictures and designs. Today we are going to look very closely at just a few of the shapes you have been using. After we talk about what some different shapes look like, we will be making the shapes with ropes of clay.

Choose the poster for one of the shapes that most students are familiar with, such as the triangle, and show it to the class.

Here is one of the shapes we have been using. Look very carefully at this shape and think about how it looks. Suppose you were going to tell someone about this shape and what it looks like. How could you describe it to them? What is one thing that you notice about this shape?

As students verbally describe shapes, they become more familiar with their characteristics and begin to recognize how various shapes are the same and different. Record students' observations and descriptions on the poster. After a number of students have shared, choose one or two comments to follow up on as a way of further describing the characteristics of a shape.

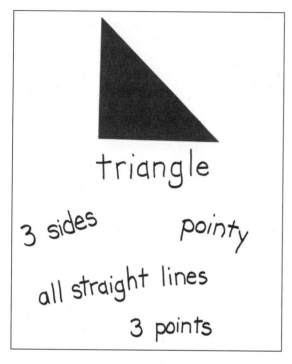

Felipe said that this shape has three sides. Let's count them together: *[point to each side]* **one, two, three. What do you notice about the sides of this shape? Are they straight? What do you notice about the corners, or angles, where the sides meet?**

Not all kindergartners will know the geometric names of shapes, but many will be able to observe and describe their attributes and characteristics. This is an important step in constructing an understanding of what makes a triangle a triangle or a square a square. You can encourage students' use of geometric vocabulary by introducing the names of shapes in this discussion. The **Dialogue Box**, Three Pointy Corners (p. 80) illustrates how one class described a triangle and how their teacher integrated the word *triangle* into their discussion.

Show students the other shape posters and ask them to describe at least one other shape in detail. It is not necessary to talk about every shape poster in this first meeting, but students should see all four posters before they begin to work with the clay.

Tell students that you will hang these shape posters around the room. During the next few days, as a class, they will talk about each remaining shape and add words of description to the posters.

Activity

Making Shapes with Clay

Note: Before beginning this activity, students should have had the chance to freely explore clay or playdough during previous Choice Times.

To demonstrate the use of clay, seat students on the floor in a circle. Choose one of the shape posters to use in your demonstration.

Show students a ball of clay and a sturdy cardboard mat. Begin by demonstrating how to roll out some long clay ropes (or "snakes," as kindergartners often call them).

Today we are going to be making different shapes with this clay. Suppose we wanted to make a triangle shape like the one on this poster. How could we use these clay ropes to make a shape that has three sides? Who has an idea?

Ask one or two volunteers to construct a triangle with the clay ropes, making their outline on the cardboard mat.

It's important to keep this demonstration as brief as possible, spending only enough time to convey the activity. While you may find a lot of different ideas to pursue as students demonstrate how to make a shape, it's better to move quickly to independent work and return to some of those ideas in subsequent group meetings.

Figuring out how to arrange the clay ropes to form a shape with three or more sides is a challenging task for many kindergarten students. To form a triangle, some students might join together three individual ropes at three corners, while others might form one long rope into a shape with three sides. Shapes made from a single rope tend to have corners that are more curved than pointed. Regardless of their method, students will be drawing on everything they have previously learned about shapes through observation and experience. See the **Teacher Note,** Making Shapes (p. 79), for more information.

Each of you will have a ball of clay to work with and a cardboard mat to build on. Your job is to make different shapes with the clay. If you need some ideas about the different kinds of shapes that you can make, you might look at one of our shape posters. Or, you can talk with your neighbors and ask them for an idea.

Distribute clay and cardboard to students. They may work individually or with a partner at tables around the classroom. The shape posters should be on display where students can easily see them.

Note: When students do Clay Shapes for Choice Time (p. 68), a set of Make-a-Shape cards will be available. These cards can be used as templates for students to build on with clay ropes. The Make-a-Shape cards are purposely *not* made available during Focus Time so that you can gather some initial information about how students approach the task of constructing shapes. That is, can they make any shapes without a direct model? If they use the shape posters as a model, how successful are they, given that they cannot build on top of the picture?

Observing the Students

Observe students as they are working on Clay Shapes.

- Are students able to make a shape using the clay? How do they describe their shape? Do they refer to the shape by name?

- Are students able to make a variety of shapes? Do they consult the shape posters, or do they seem to "just know" the shapes they want to construct?

- How do students make their shapes? Do they seem to have a plan in mind that includes the specific attributes of a shape (such as, "A triangle has three sides")? Do students make the individual sides of a shape and then attach them at the corners, or do they use one long clay rope to outline a shape?

- If students are having difficulty making a shape, note this information. You might suggest that they start with a circle, since it takes just a single rope of clay to make this very familiar shape. Show them the circle poster, or draw a circle for them to look at. If students are still having difficulty, suggest that they build a clay circle directly on top of one you have drawn on paper.

Activity

Sharing Our Clay Shapes

Plan a short sharing meeting for the end of Focus Time. About 10 minutes before the end of the session, ask students to choose (or make) a clay shape to share with the class. Any leftover clay can be rolled into a ball and placed in a designated spot.

Students bring their cardboard mats, with their clay shape, to the meeting area. When everyone is seated on the floor, ask students to look carefully at the sides of their shapes.

We are going to put our clay shapes into groups. To decide what group a shape belongs in, we'll look at the sides of the shape.

Does your shape have curved sides or round sides? If so, put it on the rug over here, to start a group of curved shapes. . . . If your shape has *three* sides, place it in a three-sided group, over here.

As needed, make groups for shapes with four sides and with *more than* four sides, and finally for shapes that don't fit any of the other groups.

When all the shapes have been grouped according to the type and number of sides, collect a few observations from students. The length of this discussion about the clay shapes will depend on students' attention and your daily schedule; as needed, you can postpone the discussion until a later time. Following are some ways you might focus the discussion:

- Look at only one group—such as the group of shapes with four sides—and make observations about the shapes in that group.
- Count the number of shapes in each group.
- Talk about what was hard about making clay shapes. Were some shapes harder to make than others?
- Discuss the shapes that students thought fit into none of the groups.
- Compare two related shapes (such as squares and rectangles, or circles and ovals). For an example of one such discussion, see the **Dialogue Box, Circles and Ovals (p. 81).**

Clay Shapes will be offered as a Choice Time activity over the following days. To be ready for that activity, students need to take apart their shapes, roll their clay together into a ball, and return the ball to an airtight container.

Focus Time Follow-Up

 **Extension**

Rope Shapes After students have had some experience making clay shapes, try making larger shapes using a large loop of rope. This can be done with the whole class or with small groups. This works best in a large, open space, such as a gymnasium or playground. Three or four students at a time may hold onto the rope and work together to form it into a shape. Encourage them to experiment with making the sides of a shape longer and shorter.

 **Choice Time**

Five Choices In addition to the Clay Shapes activity (p. 68) that is an extension of the Focus Time work, you can introduce three new activities over the next few days: Fill the Hexagons (p. 70), Build a Block (p. 73), and the computer choice Quick Images (p. 75). In addition, students can continue their work on Pattern Block Puzzles (p. 34) from the previous two investigations.

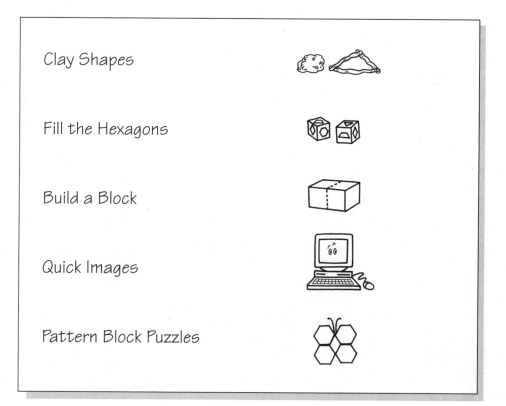

Focus Time: Clay Shapes

Choice Time

Clay Shapes

What Happens

Students continue making the outlines of shapes with ropes of clay. They are encouraged to make not only different shapes but also different examples of the same shape, such as many different triangles or rectangles. Students' work focuses on:

- becoming familiar with the attributes of 2-D shapes
- constructing 2-D shapes

Materials and Preparation

- Make available the balls of clay or playdough and the cardboard mats from Focus Time.
- Prepare five or six sets of Make-a-Shape Cards (1 set per 3–4 students) by copying pp. 178–180, preferably on card stock. Cut apart the cards and clip or rubber band each set together.

Activity

This activity needs little introduction since it is a continuation of the work students began in Focus Time. Show them a set of Make-a-Shape cards and explain that several sets of these cards are available.

One activity choice is to make shapes with ropes of clay, the way you did [yesterday]. This time you can look at these Make-a-Shape cards and try out some new shapes. Some of these shapes have sides that are curved, and some have straight sides. Some of the shapes have three sides, some have four sides, and some even have five or six sides.

If you want students to save examples of their clay shapes to share, set up a space in the classroom for saved work. As necessary, replenish and refresh the clay or playdough supply.

As students repeat this activity over time, you might direct their work in specific ways. For example:

- Suggest that they make only shapes with three sides (or shapes with four sides), making as many different examples as they can.
- In the Make-a-Shape cards, show them the shape with five sides (pentagon) or six sides (hexagon) and ask them to make different examples of shapes like this.

Observing the Students

Consider the following as you watch students making clay shapes.

- Are students able to make a shape using the clay? How do they describe their shape? Do they refer to the shape by name?
- Are students able to make a variety of shapes? Do they consult the shape posters or Make-a-Shape cards, or do they seem to "just know" the shapes they want to construct?
- How do students make their shapes? Do they seem to work from a plan that includes the specific attributes of a shape, such as "A triangle has three sides"? Do students make the individual sides of a shape and then attach them at the corners, or do they use one long clay rope to outline a shape?
- How do students use the Make-a-Shape cards? Does anyone use them as a template and make the clay shapes directly on top of the outline on the cards?

Variations

- Introduce other materials to make shapes with. If you have Geoboards and rubber bands, students can make straight-sided shapes on the boards. Students can also make straight-sided shapes with plastic drinking straws cut in varying lengths, using small pieces of clay at the corners of shapes for connectors.

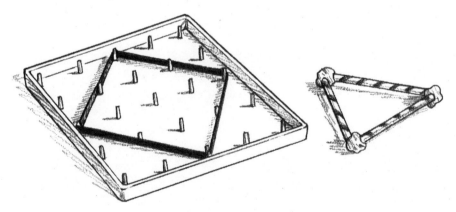

- Supply paper and pencil and ask students to draw shapes with three or four sides.

Choice Time

Fill the Hexagons

What Happens

In a game called Fill the Hexagons, players roll game cubes labeled with pictures of pattern blocks, then take the blocks rolled and use them to fill in the six hexagon outlines on the gameboard. The goal is to completely fill all six hexagons. Students' work focuses on:

- finding combinations of shapes that fill an area
- building knowledge about the relationships among pattern block shapes

Materials and Preparation

- Duplicate Student Sheet 2, Fill the Hexagons Gameboard (p. 181), preferably on card stock, and laminate if possible. You'll need 1 gameboard per player or pair of players, plus another optional copy on plain paper for each student if you plan to present the variation that involves recording (p. 72).
- Make available the pattern blocks, 1 bucket per 4–6 students.
- On small stickers that fit the faces of blank cubes, draw pictures of these four pattern block shapes in the corresponding colors: yellow hexagons, red trapezoids, blue rhombuses, green triangles. Place these stickers on blank cubes to make pattern block game cubes, each with 1 hexagon, 1 trapezoid, 2 rhombuses, and 2 triangles. Each person, pair, or group needs 2 of these game cubes to play the game.
- Make available paper pattern blocks and glue sticks if you plan to present the variation that involves recording (p. 72).

Activity

Today we're going to learn how to play a game called Fill the Hexagons. This game uses some of the pattern blocks, this gameboard, and some special game cubes that I made. *[Show students the materials as you name them.]* What do you notice about these game cubes?

Some students will recognize that the faces of the game cubes are labeled with pictures of pattern blocks. Others might notice that the orange square and the tan rhombus are missing, or that there are *two* blue rhombuses and *two* green triangles. After students have explored and commented on the game cubes, explain what players will need for the game.

You can play Fill the Hexagons by yourself, with a partner, or with a small group of people. Every person who is playing will need a gameboard. Everyone in the game can share a container of pattern blocks and two pattern block game cubes.

Play a demonstration round of the game with a volunteer, or play a round against the class as a whole, asking different students for their suggestions for placement of pattern blocks on each turn. To show students how to set up the game, give each player a gameboard and place the pattern blocks and the game cubes between you to share.

The goal of this game is to fill your whole gameboard with pattern blocks. Every hexagon needs be covered, with no empty spaces left, just the way you did the pattern block puzzles. To start the game, the first player rolls both of the game cubes. You and your partner need to decide who goes first. Who can see what Alexa rolled?. . . Alexa can take those blocks and put them anywhere on her gameboard, on any hexagon.

Players can put both blocks on the same hexagon (if possible), or they can put one block on one hexagon and the other block on another hexagon. Explain that each move is final.

When it's your turn to place the blocks, you'll need to think very carefully, because once you've put your blocks down, you can't move them on a later turn.

Choice Time: Fill the Hexagons ■ **71**

As you play this sample round, ask students to help you decide where to place your blocks and to explain their choices. Continue to play several rounds until the rules of the game are clear.

Observing the Students

Consider the following as you watch students play Fill the Hexagons.

- How easily can students place shapes? As players fill in the hexagons, do they plan ahead? Can they foresee what will happen if they place blocks in certain ways? When they place a block, do they think about what they will need to fill in the remaining space?

- How do students talk about and describe the pattern block shapes? How do they talk about and describe the spaces that remain in a partially-filled hexagon? (For example, in a hexagon outline with one trapezoid block already placed, students might hope for a roll of "the red one," "the half one," or "the trapezoid.")

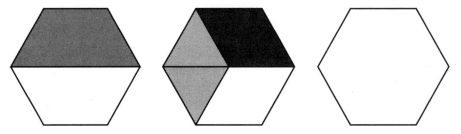

- What relationships among the shapes do students use in deciding how to place their blocks? Is there evidence that they are beginning to see and use equivalents among the shapes? For example, do they realize a space the size of the trapezoid can be filled with one red trapezoid, three green triangles, or a blue rhombus and a green triangle?

Variations

- Students could record their work with paper pattern blocks, glue sticks, and Fill the Hexagon Gameboards copied on plain paper. If students have difficulty replacing blocks with cutouts to record their solution, you might offer them two copies of the gameboard—one for building, one for recording—so that they need not take apart their solution to record.

- Some students might like to investigate how many different ways it's possible to "make" the hexagon with pattern blocks.

Choice Time

Build a Block

What Happens

Students build one of the larger Geoblock shapes by combining two or more smaller Geoblocks. Their work focuses on:

- combining smaller three-dimensional shapes to make a larger three-dimensional shape

Materials and Preparation

- Make available 2 sets of Geoblocks, each divided into half-sets. From these, pull out the largest blocks (the square prism, 8 cm by 4 cm by 4 cm)—there are two of these in a set.

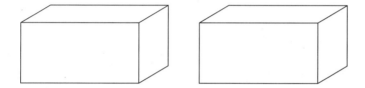

Activity

To introduce this activity, set up a display of about eight Geoblocks. Include the largest block and two 4-cm cubes that, when placed end to end, replicate the largest Geoblock. (These three blocks are shaded in the illustration.) Call attention to the largest block.

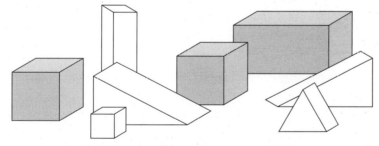

Suppose I wanted to build another large block like this one. Do you see any smaller blocks I could put together to do that?

Invite a couple of volunteers to try combining other blocks. When someone has successfully built the larger shape with smaller blocks, hold up both examples for all to see.

Choice Time: Build a Block ■ **73**

Justine used these two cubes to build a block that is exactly the same size as this large Geoblock. And that's not the only way to build a large block the same as this one. During Choice Time, when you choose Build a Block, you are going to see how many different ways you can find to put smaller blocks together to build the largest Geoblock.

Encourage students to work together in pairs on this activity. A half-set of blocks should accommodate four students at a time. Since there is no way for students to record or save their work, you will need to check periodically with the students working on this choice. Advise students that when they think they have found all the possibilities, they should save their combinations until you have had a chance to see them.

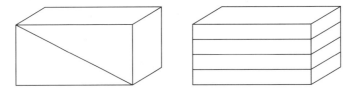

Observing the Students

Consider the following questions as you watch the students "building" Geoblocks.

- How do students put the blocks together? While this is an exploratory activity, you can learn a great deal about how students perceive three-dimensional shapes by watching how they attempt to build them.
- Do students randomly choose smaller blocks and try to build a larger one? Do they seem to recognize and specifically select blocks that are half of a larger block and put two of them together? Do they use smaller cubes to build the larger cubes?
- How do students prove to themselves that they've made an exact replica of the larger block?

Variations

- Select other blocks from the set of Geoblocks for students to build.
- Set up the largest square prism as a "base block" for a wall. Students then add on other blocks or combinations of blocks to make a wall that is the same height (4 cm) and the same width (4 cm) as the base block. This block wall can get fairly long, so provide ample space for this activity.

Choice Time

Quick Images on the Computer

What Happens

Quick Images is a computer activity that you introduce to the whole class in an off-computer version. Students briefly view an image of two pattern blocks in a particular arrangement, then, with the original image hidden, put together pattern blocks to make a copy of the image they saw. In the computer version, students work with shapes on the screen in place of actual pattern blocks. Their work focuses on:

- analyzing visual images
- describing position of and spatial relationships among objects

Materials and Preparation

- For demonstration, have a small tray or cardboard mat for holding a simple pattern block design, and a sheet of paper or cloth for covering the design.
- Distribute a cup containing three pattern blocks (hexagon, square, triangle) to each pair of students.
- For the Choice Time activity, you'll need only computers with the *Shapes* software installed.

Activity

The Quick Images activity is repeated at all grade levels of the *Investigations* curriculum. The images vary from grade to grade, and include dot patterns as well as arrangements of cubes and geometric shapes. Through Quick Images, students analyze visual images to become familiar with different kinds of numerical and geometric arrangements.

Introducing Quick Images

Bring to the meeting area a tray or cardboard mat, paper or cloth to cover it, and the cups of three pattern blocks. Give each pair of students a cup with the three pattern blocks (hexagon, square, triangle).

On this tray, I'm going to show you a design I made with pattern blocks. You'll have to look quickly and carefully at my design, because very soon I'm going to cover it up again.

When you can't see the design any more, you and your partner will try to make a copy of what you saw, using the pattern blocks in your cup.

Students won't have long to get the picture in their mind. To help them concentrate on the image of the blocks, they should not build anything or be handling their own blocks while the image is visible. You might suggest that while the image is being shown, they keep their hands in their laps and their eyes on the tray.

Beneath the paper or cloth, build an image on your tray with just two pattern blocks, such as one of the following examples.

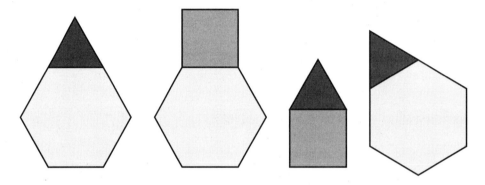

OK, hands in your laps. I'm ready to show you my pattern blocks. Get ready to look very carefully. Here it comes!

Show the design for 5 seconds and then cover it again. (As needed, adjust the amount of time you flash the image.)

Student partners now work together to make a copy of the design they saw. If some students are concerned that they cannot recall the figure exactly, assure them that they will have another chance to see the design and to revise their work.

Hands in your laps. Here comes the design again. Look carefully!

Show the image for another 5 seconds. Then let students revise their own arrangement. Ask a few volunteers to tell you what they saw and how they remembered what to build with their blocks. Reassure students that you know it can be difficult when they've seen your blocks only briefly. After this discussion, show your design one more time, leaving it visible for further revision. Encourage students to talk about any revisions they make in their work.

You might repeat the activity once or twice, depending on students' interest and on your time constraints. Explain that during Choice Time, they'll be doing this same type of activity on the computer. Introduce a small group to

Quick Images on the computer while others cycle through the other Choice Time activities.

Quick Images on the Computer

Begin by showing students how to open the *Shapes* software and choose the Quick Images activity. Read aloud the directions in the box that appears, and show them how to click on OK.

Demonstrate that clicking on the eyes gives the viewer a short peek at an arrangement of pattern blocks. Students try to copy the design they saw briefly by dragging shapes from the Shape bar into the window on the left. Ask students to help you build a copy of the first image and to explain to you how they are remembering it.

Show students that clicking on the eyes a second time will give them another look at the image—this time slightly longer—so that they can revise their work. They can continue to click on the eyes as necessary, for up to nine (successively longer) glimpses. After nine glimpses, the image remains visible, and students can build a copy directly from the picture. The little number next to the eyes tells them how many "peeks" they have taken. Be alert to and discourage any competition that arises over the number of peeks it takes different students to build the image.

Talk with students about how they know when they have built an exact copy of the image that was flashed. While this may be obvious to adults, many young students will not perceive a difference in orientation as making a different shape. They tend to look first at having the right blocks in their design, and some may need a reminder to consider how they are positioned.

Every time Quick Images opens, the software presents image Number 1. Students use the Number menu to select the image they want. Demonstrate how to choose a new image by clicking on Number in the menu bar and dragging the cursor down another number.

Note: Numbers 9 and 10 are more difficult images and should be used for kindergartners who are ready for more challenge.

Each time students move to close the activity or choose a new image, a box will appear, asking if they want to save the work they've done on the current image. They may want to save, for example, if time is up and they aren't finished with their work, or if they have finished and want to show you that they successfully copied the image. If you want students to save their work to a disk, explain this process to them now (refer to p. 153 as needed). Otherwise, explain that they may click on Don't Save to continue with the next image.

Observing the Students

Consider the following questions as you watch students working on Quick Images at the computer.

- Do students understand how to use the computer and its tools to work on this task?
- How do students approach the task? Do they have a strategy that helps them remember the image? For example, after the first peek, some students try to remember and retrieve all the blocks seen, and then use the second peek to concentrate on how to arrange and orient those blocks. Other students try to remember the blocks in a particular order, remembering the color, shape, or size of blocks. For example, for image Number 2, they might chant to themselves "triangle, then hexagon, then trapezoid," or "green, yellow, red," or "little, big, middle." Do they associate the image with some real object to help them recreate it? For example, image Number 1 might remind them of "a head with a hat."

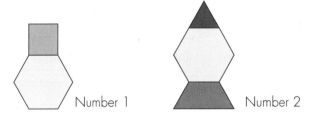

Number 1   Number 2

Holding on to this much information will be hard for many kindergartners. Encourage students to talk to each other about the strategies they are using to remember the images, but reassure students for whom it takes many peeks. Remind them that if they persist at this activity, over time they will get better at keeping the image in their mind.

- How are students using the tools in *Shapes*? Are they becoming more comfortable and competent with tools such as Turn, Flip, and Duplicate?

Variation

- Continue using actual pattern blocks to do Quick Images with the whole group. You could also expand this activity to include images that are arrangements of counters, cubes, or dot patterns drawn on the board.

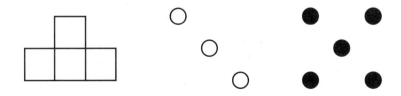

Making Shapes

Teacher Note

One of the ways students learn about shapes is by *constructing* them. Just looking at a shape, whether a triangle or a circle, is a different experience from trying to make that shape. In order to draw a triangle or make one with clay, students have to think about all the parts of the shape and how they go together. Very young children, when trying to draw a triangle, may include straight sides and some corners but have a great deal of difficulty getting the right number of sides and angles in the right positions. By kindergarten age, students begin to notice how many sides shapes have, whether the sides are straight or curvy, and how "slanty" the angles look.

Drawing or constructing shapes gives students an opportunity to think more carefully about the attributes of the shapes they are making. Drawing or outlining shapes with ropes of clay involves students in moving around the edges of the shapes and feeling the straight sides, the curves, and the angles. Their visual and tactile impressions are combined as they work to make the shape look right.

Putting together shapes to build other shapes also helps students develop their knowledge about the characteristics of different shapes. When students use pattern blocks to fill in an outline, they begin to look at angles, the lengths of sides, and what can fit in a particular space. For example, when there is a certain space they want to fill in, they might look at the corners of a pattern block piece, judging the size of the angle to see if it will fit in the space: Will a *square* corner fit? a smaller angle? a larger one?

Similarly, building with Geoblocks offers many opportunities to notice the characteristics of shapes. In the activity Build a Block, students put together several smaller Geoblocks to "build" a particular larger Geoblock. As they do this, they are matching faces of Geoblocks, paying attention to lengths of sides, thinking about the sizes of blocks in relation to each other, visualizing what a block might look like cut into two or more parts, and noticing whether sides come together in "square corners" or "slanty" ones.

As students draw, make clay shapes, and build with blocks, talking about what they are doing is an important part of the learning process. In class discussions about making shapes, students not only learn to describe what they are noticing about shapes, but also encounter new ideas that may deepen or challenge their own, and hear words from their classmates and teacher that can expand the language they use to describe shapes.

DIALOGUE BOX

Three Pointy Corners

In preparation for the Clay Shapes activity, this kindergarten class is describing one of the shape posters, the triangle. The teacher first solicits observations about the shapes from the students, then uses those observations to build a description of a triangle.

Suppose you were going to tell someone about this shape and what it looks like. How could you describe it? What is one thing that you notice about this shape?

Kadim: It has three lines.

Charlotte: The lines are black.

Henry: It has points.

Justine: Those points are corners, I think. It has three pointy corners.

Miyuki: It's a triangle. It sort of looks like a hat, too.

Kadim said that this shape has three lines. Can anyone say more about those lines?

Henry: Two of them aren't straight, but the one on the bottom is straight.

Maddy: The lines *are too* straight. They *have* to be straight.

Can you show me what you mean by "not straight," Henry?

Henry: *[Coming up to the shape poster, he traces the two angled sides of the triangle.]* See, this side and this side are not up straight—they go in. Kind of like an A. And this line on the bottom is straight.

So you were noticing that two of the sides of this shape were not straight up and down, but instead they were slanted like the letter A.

Maddy: Oh, I thought you meant that they were *curvy,* like the letter O!

Miyuki called this shape a triangle. Other people noticed that the shape has three lines or sides that are straight, and that the sides slant in, and that the triangle has three pointy corners. All of these are important parts of the triangle shape. Did anyone notice any other things about this shape?

The teacher is pleased that students are beginning to pay attention to important characteristics of shapes. Even though they don't always use mathematical terms (for example, some say "lines" for *sides* and "corners" for *angles*), they are clearly communicating what they mean. When one student uses the term *triangle* to describe the shape, the teacher acknowledges the word and incorporates it into the discussion. Without expecting everyone to use the correct geometric names, the teacher continues to use such words so that students can begin to build their knowledge of the shapes these words describe.

DIALOGUE BOX

Circles and Ovals

These kindergartners have been working with ropes of clay to make shapes—squares, rectangles, triangles, circles, hexagons, and others. Since quite a few students have been talking among themselves about how ovals are different from circles, the teacher decides to focus the class discussion on this topic. Students who have made circles and ovals are asked to bring them to the group meeting on their cardboard mats.

So, when you were making a circle, what did you have to think about?

Ravi: It's round and curved.

Ida: It has round edges.

Ayesha: It's like this *[shows her clay circle]*.

Suppose I couldn't see Ayesha's circle. How could I describe what it's like?

Thomas: It has two half circles.

Renata: It has no sides—it curves around.

Miyuki: Circles are like C's *[the letter]*, but you keep going and close it up.

Brendan: Half circles are like D's, but no line going down. See *[showing his mat]*, you put two D's together, like this, but then you'd have to take out the line.

I noticed some of you were making a different kind of curved shape. How about you, Xing-Qi? What is yours?

Xing-Qi: It's an oval *[shows his clay shape]*.

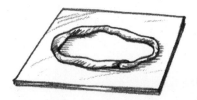

So what can you tell me about an oval?

Tess: They're circles.

Carlo: Ovals and circles are round, but the oval is longer.

Luke: The oval is stretched out like an egg and not as round as the circle.

Charlotte: They're both round, though.

Gabriela: They're like circles, but they're tall.

So, some of you think an oval and a circle are the same in some ways. And some of you see some differences. When you were making them, did you notice that you had to do some things the same with the clay?

Tess: You had to keep winding it. It couldn't go straight, because then it would be a square or something.

Carlo: Your hand gets tired because you have to keep turning, turning all the time.

Renata: But with the circle, it's really hard because you're always trying to get it even all around, but with the oval, it can be kind of squooshed. It doesn't have to be even.

What do you think Renata means by "even"? Does a circle have to be even?

Alexa: It's like when you tell us to sit in a nice circle. We're all even around the edges.

Kylie: It has to keep curving and curving. It can't get all slanty like on the bottom of the oval.

While they compared and contrasted two shapes that have some common attributes and some differences, students could begin to articulate the characteristics of these common shapes. Describing how they moved the clay as they made the shapes provided another way for students to think about what makes a circle a circle or an oval an oval.

INVESTIGATION 5

2-D Faces on 3-D Blocks

Focus Time

A Close Look at Geoblocks (p. 84)

Students discuss the attributes of Geoblocks while looking closely at these 3-D shapes. They examine the faces (sides) of the individual Geoblocks and search for two blocks that have matching or congruent faces.

Choice Time

Matching Faces (p. 88)

As a follow-up to the group work during Focus Time, students work with a partner or small group to find pairs of Geoblocks that have congruent or matching faces.

Geoblock Match-Up (p. 90)

As students play the game Geoblock Match-Up, they match faces of Geoblocks to outlines on a gameboard.

Planning Pictures on the Computer (p. 92)

After making a simple picture with actual pattern blocks, students use the *Shapes* software to replicate their creation on the computer.

Continuing from Investigation 4
Clay Shapes (p. 68)
Fill the Hexagons (p. 70)

Mathematical Emphasis

- Observing and describing attributes of 3-D shapes
- Looking at 3-D objects as wholes and as having parts
- Observing similarities and differences between the faces of different 3-D shapes
- Matching a 3-D block to a 2-D outline of one of the block faces

Teacher Support

Dialogue Box
Mine Looks Like a Ramp (p. 95)

INVESTIGATION 5

What to Plan Ahead of Time

Focus Time Materials

A Close Look at Geoblocks
- Geoblock sets to share

Choice Time Materials

Matching Faces
- Geoblocks: 2 sets, divided equally into half-sets

Geoblock Match-Up
- Half-sets of Geoblocks
- Geoblock Match-Up Gameboards (pp. 182–187): 1–2 of each, copied on card stock and, if possible, laminated

Planning Pictures on the Computer
- Computers with *Shapes* software installed
- Pattern blocks
- Trays or sturdy cardboard mats (optional)

Clay Shapes
- Clay or playdough and cardboard mats from Investigation 4
- Make-a-Shape cards: 5–6 sets from Investigation 4

Fill the Hexagons
- Student Sheet 2, Fill the Hexagons Gameboards: class supply from Investigation 4
- Pattern block game cubes from Investigation 4
- Pattern blocks: 1 bucket per 4–6 students

Focus Time

A Close Look at Geoblocks

What Happens

Students describe general characteristics of the Geoblocks, looking at both the overall 3-D shapes and at specific parts of each shape. Then, looking very closely at the faces (sides) of individual Geoblocks, they search for two blocks that have matching (or congruent) faces. Their work focuses on:

- observing and describing attributes of 3-D shapes
- matching faces of 3-D shapes

Materials and Preparation

- Have a half-set of Geoblocks available for use during Choice Time.

Activity

A Close Look at Geoblocks

Pass around the container of Geoblocks. Ask each student to choose one block and to look at it very carefully.

If you were going to describe your block to someone, what is one thing you could say about the block you are holding?

Students will likely mention a wide variety of things about the Geoblocks. For example:

- They may point out faces of the blocks that are particular shapes—squares, rectangles, triangles, or diamonds.
- They may count the number of "sides" (faces), "lines" (edges), or "points" (corners) on their block.
- They may tell what the block or part of the block "looks like"—a box, a present, a number cube, a ramp, a roof, and so forth.
- They may give more general descriptions of their block—it's made of wood, it's smooth, it's pointy. Or they may describe what they might do or build with it: "I have a block I'd use to make a bridge [building, tower, school]."

Encourage students to think about each observation that is made, and to consider whether their own blocks also match that description. For example, if one student says "My block looks like a ramp," ask others who think their block also looks like a ramp to raise their hand or hold up their block. This gives everyone a chance to share without going around the group one by one. See the **Dialogue Box,** Mine Looks Like a Ramp (p. 95), for this discussion as it occurred in one kindergarten class.

After students have described their block, ask them to compare their block to the one their neighbor has.

I'd like you to turn to a person who is sitting near you and compare your two blocks. How are they the same? How are they different? You can think about some of the descriptions that we just talked about, and you may also notice new things.

After some discussion, focus attention on the faces of the Geoblocks. Suggest that pairs of students see if their two blocks have faces that could be put together so that they match exactly. Remind them of the way they matched faces of blocks in the Build a Block activity. Show an example of what you mean, choosing two blocks with congruent faces and placing those two blocks together.

If you and your partner have blocks with a matching face, put those two blocks together and place them in the center of our circle where everyone can see them.

When all the pairs with matching faces have been placed, call attention to those pairs with two blocks that are exactly the same (Example 1), and those pairs with two blocks of different shapes but still sharing a congruent face (Example 2).

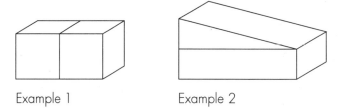

Example 1   Example 2

If there are no examples of different-shaped blocks with matching faces, find one example in the Geoblock set to share with the class.

Here are two blocks with different shapes, but I can see a way to put the faces together so they match exactly. Can you?

Initially many students will match only blocks that are exact duplicates. Some students will find common attributes of blocks, saying, for example, "They both have a triangle side" (Example 3), but not recognizing that the triangles are different sizes or different types, not congruent, and therefore cannot be matched exactly.

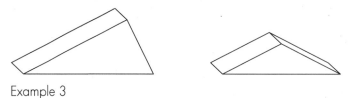

Example 3

Explain that during Choice Time, in the activity Matching Faces, students will look through the class sets of Geoblocks to find pairs with faces that can be matched together. (If there is time, you may want to proceed to that activity now.)

Focus Time Follow-Up

 **Choice Time**

Five Choices In addition to the activity Matching Faces (p. 88), which is a continuation of the Focus Time work, you can introduce another activity that involves looking at the faces of the Geoblocks: Geoblock Match-Up (p. 90). In this game, students match the faces of Geoblocks to outlines of six shapes on gameboards. For classes using computers, the *Shapes* activity Planning Pictures (p. 92) can be introduced to small groups of students. Students may also continue with two activities from Investigation 4, Clay Shapes (p. 68) and Fill the Hexagons (p. 70).

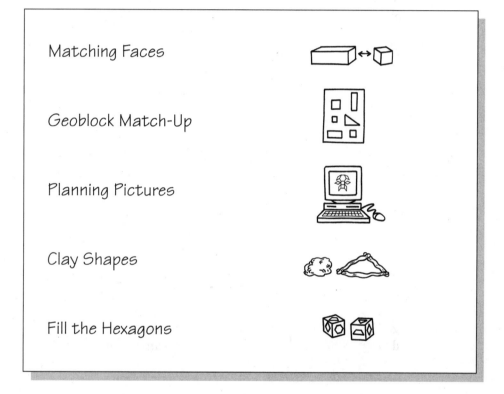

Choice Time

Matching Faces

What Happens

Students work with a partner or small group to find pairs of Geoblocks that have congruent faces. Their work focuses on:

- observing similarities and differences between the faces of different 3-D shapes
- observing similarities and differences between the different faces of a single 3-D shape (for example, noticing that triangular prisms have some rectangular faces, even though the overall shape seems to be "triangular")
- developing vocabulary to describe the different faces (square, rectangle, and triangle) of 3-D shapes

Materials and Preparation

- Have your half-sets of Geoblocks set out in containers for small groups to share. Note that these sets also need to be available for the Choice Time activity, Geoblock Match-Up.

Activity

Because this activity simply extends the Focus Time work, it needs only a brief introduction.

In the Matching Faces activity, you and a partner will work together to find two Geoblocks that have a matching face. When you have found two blocks, match the faces together and place them on the table. Then look for more pairs with matching faces. Try to find as many pairs of blocks as you can.

Periodically check in with students to observe their pairs of blocks. (Since the Geoblocks are needed for two activities, it will be difficult to save their work for long.)

  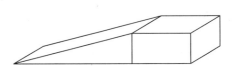

Observing the Students

Consider the following questions as you watch students working on Matching Faces.

- Do students find matching block faces easily or with difficulty?
- Do students notice that the triangular prisms have some rectangular faces, even though the overall shape seems to be "triangular"?
- Do students keep in mind all the faces on a block as they work? That is, when a particular face doesn't match one face of another block, do they check to see if it matches any of the other faces on that block?
- Do students differentiate between different sizes of the same shape?
- What vocabulary do students use to describe the 3-D shapes? Do they use the names of familiar 2-D shapes to describe the 3-D shape, for example, calling the cube a *square* or the triangular prism a *triangle?*

Variations

- Students who are familiar with the game Go Fish can play a version with Geoblocks. Each player takes six blocks from the Geoblock set and makes any "matches" possible within those six blocks. (A "match" refers to matching faces.) Players then take turns "fishing" in the tub to find a match for their remaining blocks. This fishing can be done with eyes open or closed, but players should agree on the rules beforehand. Play continues until all players have found a match for each of their six Geoblocks.

- Choose five or six matched pairs of blocks (that is, blocks with one face matching). Players separate the matched pairs, mix up the blocks, and spread them out. They then take turns looking for matches—not necessarily the original pair, but any matching pair. The goal is for the players to cooperatively match up all the blocks, leaving as few leftover, unmatched blocks as possible at the end. If they are unable to use all the blocks, students may want to take apart and rematch some blocks until all are used.

Choice Time

Geoblock Match-Up

What Happens

Using any of six Geoblock Match-Up gameboards, students look for Geoblocks with faces that match the outlines on their board. Their work focuses on:

- developing vocabulary to describe 3-D shapes
- looking at 3-D objects both as wholes and as having parts
- matching a 3-D block to a 2-D outline of one of the block faces
- describing and comparing faces of 3-D shapes

Materials and Preparation

- Prepare Geoblock Match-Up Gameboards 1–6 by copying them on card stock. Laminate if possible. Each set of six boards will allow six children to play the game at a time; you may want to make two or more full sets.
- Students will need access to the half-sets of Geoblocks also being used for the Choice Time activity Matching Faces.

Activity

To introduce this game, have at hand some Geoblocks and the Geoblock Match-Up Gameboards. Remind students of the work they have been doing with Geoblock faces, and show them a few pairs of blocks with matching faces.

You have been looking very closely at the sides, or the *faces*, of the Geoblocks. You know how to look for two Geoblocks with matching faces—when you put those faces together, they match exactly. In this new game, called Geoblock Match-Up, you are going to look for another kind of match. You'll have a gameboard like this *[display one]*, and you'll try to find a block that exactly matches each outline on your gameboard. You need to find blocks for all six outlines.

There are six different gameboards; each player needs just one. The set of Geoblocks should be within easy reach of all players.

Player take turns reaching into the container and choosing one block. If one of the block's faces matches an outline on the player's board, he or she places that block on top of the outline and the next player takes a turn.

■ *Investigation 5: 2-D Faces on 3-D Blocks*

A player who cannot find a match puts the block back in the container, and the turn passes to the next player. The object is for each player to find a match for all six outlines on his or her gameboard.

When students first play this game, their focus tends to be narrow. Often you'll see them looking for one specific block to match one specific shape on their board. As they become more familiar with how the blocks are constructed and how they relate to each other, students begin to see more readily that one block may fit more than one outline, and that different blocks with congruent faces can match the same outline.

Although the object of the game is for a player to be the first to fill all the outlines on his or her gameboard, this game usually ends up being more cooperative than competitive. Students often help one another "see" faces that match outlines, and may even play independently of one another, as if doing a solitaire puzzle rather than playing a game. You could also encourage students to work together on a single gameboard, or to work cooperatively to fill all the outlines on two or more gameboards.

Observing the Students

Consider the following questions as you observe students playing Geoblock Match-Up.

- Do students find block faces that match the outlines easily or with difficulty?
- Do students keep in mind all the outline shapes on their board as they work? For example, when a block's face doesn't match one outline, do they check to see if it matches any others?
- Do students notice that different blocks will match the same outline?
- Can students differentiate among varying sizes of the same shape, such as big triangle, smaller triangle?
- What vocabulary do students use to describe squares, rectangles, and triangles? Does their language describe differences between the shapes? For example, do they talk about shapes that are *thick* and *thin,* or *tall* and *short?*
- Do students notice that the triangular prisms have some rectangular faces?

Choice Time

Planning Pictures on the Computer

What Happens

Pairs of students work together to plan a picture made with 5–12 pattern blocks. They then work together at the computer, using the *Shapes* software to replicate their block design. Their work focuses on:

- using pattern blocks to make a design or picture
- using *Shapes* to replicate a pattern block design or picture

Materials and Preparation

- Students will need access to computers with *Shapes* software installed.
- If there is enough space, set out 20 or so pattern blocks at each computer. If space is too tight, set up another work space with the pattern blocks and let students use trays or sturdy cardboard mats to transport their pattern block creations to the computer.

Activity

Explain to the whole class that there is now a new computer activity, using both actual pattern blocks *and* the *Shapes* software. Tell them you will introduce this choice to small groups of students at the computer, while the rest of the class participates in other Choice Time activities. Reassure students that everyone will get a chance to work on this activity.

Today I'm going to introduce a new computer choice, called Planning Pictures. It uses the *Shapes* software and pattern blocks. This time, you and a partner will create a picture or design using just a few pattern blocks. Then your task is to make exactly the same picture or design on the computer.

At the Computer To introduce this activity to a small group of students, design a simple picture with them, using several pattern blocks. Demonstrate the process of copying this creation on the computer. Open *Shapes;* open the activity Planning Pictures; read the directions, and click on OK. Then work together, using the tools in *Shapes* to replicate your pattern block picture.

Pattern block pictures should use from 5 to 12 pattern blocks. You can adjust this number as necessary for particular pairs, but keep in mind that pictures with large numbers of pattern blocks will take a lot of time to copy and may be frustrating for students. Depending on your setup, pairs can design their picture at the computer and begin with *Shapes* as soon as they are ready, or they can work with the pattern blocks elsewhere and use a tray or cardboard mat to transport their picture to the computer.

Note: You may wonder why students are asked to build a picture with the regular pattern blocks before using *Shapes*. This approach requires students to build a mental image that they then "rebuild" on the computer. If students use only the computer to make a picture, they tend to leave the shapes in their initial orientations, as presented by the computer. Using regular pattern blocks, they naturally turn them this way and that. Thus, copying a picture requires that they make all those motions explicit in the *Shapes* software. For more on this, review the **Teacher Note**, Why Use Pattern Blocks on the Computer? (p. 37).

With their previous work in *Shapes,* students should by now be familiar with the basic uses of the Arrow, Flip, Turn, and Erase tools. If (or as) students indicate the need for them, introduce other available tools and their uses.

 The Duplicate tool makes copies of shapes, either single shapes or glued-together groups of shapes.

 With the Arrow tool, students can select several shapes—useful if they want to apply a tool (such as Flip) or a menu command (such as Bring to Front) to all those shapes at the same time.

 The Magnification tools make shapes bigger or smaller. *Shapes that are different sizes will not snap to each other.*

 The Glue tool glues several shapes together into a "group." This new composite shape can be slid, turned, and flipped as a unit.

 The Hammer breaks apart a group made with the Glue tool. Clicking with the hammer on any shape in a group breaks it apart so the shapes can be moved individually.

 The Freeze tool "freezes" shapes so they can't be moved or erased. Students can then manipulate adjacent shapes without accidentally moving or erasing the frozen shapes.

 The Unfreeze tool reverses the freezing step, so that frozen shapes can once again be moved or erased.

Observing the Students

Consider the following questions as you observe students at work on Planning Pictures.

- What kinds of designs or pictures do students create with the pattern blocks? Do they consider what they know about making pictures on the computer as they create their design? For example, do they change the orientation of an actual block, just so they don't have to turn or flip it when recreating the design on the computer?

- How comfortable are students with selecting, moving, turning, and flipping pattern blocks, both on and off the computer? On the computer, do they accurately choose blocks and then motions for those blocks to match the picture they are copying?

- Do students recognize when there is a discrepancy between their original picture and their copy? How do they handle this kind of discrepancy? Do they adjust their computer design? their pattern block design?

Toward the end of each Choice Time session, give a 5-minute warning to students who are working on the computer. This allows time for students who have not finished to save their design. Students who *have* finished might also save or print out their creation. To share students' work, display their printouts or occasionally leave students' work visible on the screens for others to see.

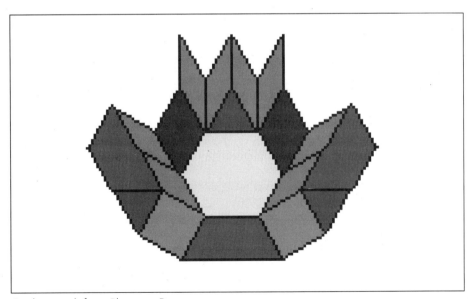

Student work from Planning Pictures

Mine Looks Like a Ramp

In this class, students have been doing free exploration with the Geoblocks. Now the teacher introduces the Focus Time activity, A Close Look at Geoblocks, and encourages the students to describe the attributes of their Geoblocks.

Today we're not going to be building with the Geoblocks. Instead, we're going to be looking very carefully at them and describing what they look like. I'm going to walk around our circle with the tub. When I come to you, take one block. Then use the best skills of observation that you have. Look very carefully. What do you notice? *[The teacher circulates with the tub of blocks.]* **Take a minute to look at your block very carefully. When you're ready to tell us about it, raise your hand.**

Tarik: I noticed mine looks like a ramp that a wheelchair would go up or down on.

Tarik says his looks like a ramp for wheelchairs to go up and down. Raise your hand if you have a block that looks like a ramp. *[Several hands go up.]*

Justine: When you look at the side of it, you see a triangle.

Justine, can you point to the face of your block that looks like a triangle? *[The girl does.]* **The rest of you, look on your blocks and see if you see any triangles. Raise your hand if you do.** *[Another show of hands.]*

Tiana: I saw on the bottom it's a square, and on top it kind of looks like a triangle, but kind of stretched out long. *[She has a pyramid.]*

If your block has both a square *and* triangles, raise your hand.

Felipe: My block is very smooth on the sides and it's pointy on the corners. I see one, two, three, four, five, six corners and four sides.

What's a corner? *[Felipe points, and the teacher holds up another block to demonstrate the corners he is pointing to.]* **Can everyone put their finger on a corner? How many corners?**

Felipe: Umm, six. *[He has a triangular prism.]*

There's lots of counting while different students count their corners.

Henry: I have eight points.

How many counted eight corners? *[Several hands go up.]* **six corners?** *[Lots of hands.]*

Shanique: I got . . . umm . . . four. Or, is this a corner right here? *[She points to the top of a triangular prism.]*

Yes, that's a corner.

Shanique: Then six.

Brendan: I got 13. I don't think I counted right.

Thomas: My block looks like a jump for skateboarders.

Like the ramps people saw earlier.

Ayesha: I noticed that my block's like a piece of chalk, when you keep on using it on the side and it turns flat [on the bottom].

Oscar: My shape kind of looks like that too.

Oscar, I'd like you to keep that in mind because we are going to play a matching game with the blocks today. Right now I'd like you to look very carefully at your neighbor's block and see if there are any things that are the same about your two blocks.

Teacher Note: About Choice Time

Choice Time is an opportunity for students to work on a variety of activities that focus on similar mathematical content. In the kindergarten *Investigations* curriculum, Choice Time is a regular feature that follows each whole-group Focus Time. The activities in Choice Time are not sequential; as students move among them, they continually revisit the important concepts and ideas they are learning in that unit. Many Choice Time activities are designed with the intent that students will work on them more than once. As they play a game a second or third time, use a material over and over again, or solve many similar problems, students are able to refine their strategies, see a variety of approaches, and bring new knowledge to familiar experiences.

Scheduling Choice Time

Scheduling of the suggested Choice Time activities will depend on the structure of your classroom day. Many kindergarten teachers already have some type of "activity time" built into their daily schedule, and the Choice Time activities described in each investigation can easily be presented during these times. Some classrooms have a designated math time once a day or at least three or four times a week. In these cases you might spend one or two math times on a Focus Time activity, followed by five to seven days of Choice Time during math, with students choosing among three or four activities. New activities can be added every few days.

Setting Up the Choices

Many kindergarten teachers set up the Choice Time activities at centers or stations around the room. At each center, students will find the materials needed to complete the activity. Other teachers prefer to keep materials stored in a central location; students then take the materials they need to a designated workplace. In either case, materials should be readily accessible. When choosing an arrangement, you may need to experiment with a few different structures before finding the setup that works best for you and your students.

We suggest that you limit the number of students doing a Choice Time activity at any one time. In many cases, the quantity of materials available establishes the limit. Even if this is not the case, limiting the number is advisable because it gives students the opportunity to work in smaller groups. It also gives them a chance to do some choices more than once.

In the quantity of materials specified for each Choice Time activity, "per pair" refers to the number of students who will be doing that activity at the same time (usually not the entire class). You can plan the actual quantity needed for your class once you decide how many other activities will be available at the same time.

Many kindergarten teachers use some form of chart or Choice Board that tells which activities are available and for how many students. This organizer can be as simple as a list of the activities on chart paper, each activity identified with a little sketch. Ideas for pictures to help identify each different activity are found with the blackline masters for each kindergarten unit.

In some classrooms, teachers make permanent Choice Boards by attaching small hooks or Velcro strips onto a large board or heavy cardboard. The choices are written on individual strips and hung on the board. Next to each choice are additional hooks or Velcro pieces that indicate the number of students who can be working at that activity. Students each have a small name tag that they are responsible for moving around the Choice Board as they proceed from activity to activity.

Introducing New Choices

Choice Time activities are suggested at the end of each Focus Time. Plan to introduce these gradually, over a few days, rather than all at once on the same day. Often two or three of the choices will be familiar to students already, either because they are a direct extension of the Focus Time activity or because they are continuing from a previous investigation. On the first day of

Teacher Note continued

Choice Time, you might begin with the familiar activities and perhaps introduce one new activity. On subsequent days, one or two new activities can be introduced to students as you get them started on their Choice Time work. Most teachers find it both more efficient and more effective to introduce activities to the whole class at once.

Managing Choice Time

During the first weeks of Choice Time, you will need to take an active role in helping students learn the routine, your expectations, and how to plan what they do. We do not recommend organizing students into groups and circulating the groups every 15–20 minutes. For some students, this set time may be too long to spend at an activity; others may have only begun to explore the activity when it's time to move on. Instead, we recommend that you support students in making their own decisions about the activities they do. Making choices, planning their time, and taking responsibility for their own learning are important aspects of the school experience. If some students return to the same activity over and over again without trying other choices, suggest that they make a different first choice and then do the favorite activity as a second choice.

When a new choice is introduced, many students want to do it first. Initially you will need to give lots of reassurance that every student will have the chance to try each choice.

As students become more familiar with the Choice Time routine and the classroom structure, they will come to trust that activities are available for many days at a time.

For some activities, students will have a "product" to save or share. Some teachers provide folders where students can keep their work for each unit. Other teachers collect students' work in a central spot, then file it in individual student folders. In kindergarten, many of the products will not be on tidy sheets of paper. Instead, students will be making constructions out of pattern blocks and interlocking cubes, drawing graphs on large pieces of drawing paper, and creating patterns on long strips of paper.

Continued on next page

Teacher Note *continued*

For some activities, such as the counting games they play again and again, there may be no actual "product." For this reason, some teachers take photographs or jot down short anecdotal observations to record the work of their kindergartners.

During the second half of the year, or when students seem very comfortable with Choice Time, you might consider asking them to keep track of the choices they have completed. This can be set up in one of these ways:

- Students each have a blank sheet of paper. When they have completed an activity, they record its name or picture on the paper.

- Post a sheet of lined paper at each station, or a sheet for each choice at the front of the room. At the top of the sheet, write the name of one activity with the corresponding picture. When students have completed an activity, they print their name on the appropriate sheet.

Some teachers keep a date stamp at each station or at the front of the room, making it easy for students to record the date as well. As they complete each choice, students place in a designated spot any work they have done during that activity.

In addition to learning about how to make choices and how to work productively on their own, students should be expected to take responsibility for cleaning up and returning materials to their appropriate storage locations. This requires a certain amount of organization on the part of the teacher—making sure storage bins are clearly labeled, and offering some instruction about how to clean up and how to care for the various materials. Giving students a "5 minutes until cleanup" warning before the end of any Choice Time session allows students to finish what they are working on and prepare for the upcoming transition.

At the end of a Choice Time, spend a few minutes discussing with students what went smoothly, what sorts of issues arose and how they were resolved, and what students enjoyed or found difficult. Encourage students to be involved in the process of finding solutions to problems that come up in the classroom. In doing so, they take some responsibility for their own behavior and become involved with establishing classroom policies.

Observing and Working with Students

During the initial weeks of Choice Time, much of your time will be spent in classroom management, circulating around the room, helping students get settled into activities, and monitoring the process of making choices and moving from one activity to another. Once routines are familiar and well established, however, students will become more independent and responsible for their own work. At this point, you will have time to observe and listen to students while they work. You might plan to meet with individual students, pairs, or small groups that need help; you might focus on students you haven't had a chance to observe before; or you might do individual assessments. The section About Assessment (p. I-8) explains the importance of this type of observation in the kindergarten curriculum and offers some suggestions for recording and using your observations.

Materials as Tools for Learning

Teacher Note

Concrete materials are used throughout the *Investigations* curriculum as tools for learning. Students of all ages benefit from being able to use materials to model problems and explain their thinking.

The more available materials are, the more likely students are to use them. Having materials available means that they are readily accessible and that students are allowed to make decisions about which tools to use and when to use them. In much the same way that you choose the best tool to use for certain projects or tasks, students also should be encouraged to think about which material best meets their needs. To store manipulatives where they are easily accessible to the class, many teachers use plastic tubs or shoe boxes arranged on a bookshelf or along a windowsill. This storage can hold pattern blocks, Geoblocks, interlocking cubes, square tiles, counters such as buttons or bread tabs, and paper for student use.

It is important to encourage all students to use materials. If manipulatives are used only when someone is having difficulty, students can get the mistaken idea that using materials is a less sophisticated and less valued way of solving a problem. Encourage students to talk about how they used certain materials. They should see how different people, including the teacher, use a variety of materials in solving the same problem.

Introducing a New Material: Free Exploration
Students need time to explore a new material before using it in structured activities. By freely exploring a material, students will discover many of its important characteristics and will have some understanding of when it might make sense to use it. Although some free exploration should be done during regular math time, many teachers make materials available to students during free times or before or after school. Each new material may present particular issues that you will want to discuss with your students. For example, to head off the natural tendency of some children to make guns with the interlocking cubes, you might establish a rule of "no weapons in the classroom." Some students like to build very tall structures with the Geoblocks. You may want to specify certain places where tall structures can be made—for example, on the floor in a particular corner—so that when they come crashing down, they are contained in that area.

Establishing Routines for Using Materials
Establish clear expectations about how materials will be used and cared for. Consider asking the students to suggest rules for how materials should and should not be used; they are often more attentive to rules and policies that they have helped create.

Initially you may need to place buckets of materials close to students as they work. Gradually, students should be expected to decide what they need and get materials on their own.

Plan a cleanup routine at the end of each class. Making an announcement a few minutes before the end of a work period helps prepare students for the transition that is about to occur. You can then give students several minutes to return materials to their containers and double-check the floor for any stray materials. Most teachers find that establishing routines for using and caring for materials at the beginning of the year is well worth the time and effort.

Teacher Note: Encouraging Students to Think, Reason, and Share Ideas

Students need to take an active role in mathematics class. They must do more than get correct answers; they must think critically about their ideas, give reasons for their answers, and communicate their ideas to others. Reflecting on one's thinking and learning is a challenge for all learners, but even the youngest students can begin to engage in this important aspect of mathematics learning.

Teachers can help students develop their thinking and reasoning. By asking "How did you find your answer?" or "How do you know?" you encourage students to explain their thinking. If these questions evoke answers such as "I just knew it" or no response at all, you might reflect back something you observed as they were working, such as "I noticed that you made two towers of cubes when you were solving this problem." This gives students a concrete example they can use in thinking about and explaining how they found their solutions.

You can also encourage students to record their ideas by building concrete models, drawing pictures, or starting to print numbers and words. Just as we encourage students to draw pictures that tell stories before they are fluent readers and writers, we should help them see that their mathematical ideas can be recorded on paper. When students are called on to share this work with the class, they learn that their mathematical thinking is valued and they develop confidence in their ideas. Because communicating about ideas is central to learning mathematics, it is important to establish the expectation that students will describe their work and their thinking, even in kindergarten.

There is a delicate balance between the value of having students share their thinking and the ability of 5- and 6-year-olds to sit and listen for extended periods of time. In kindergarten classrooms where we observed the best discussions, talking about mathematical ideas and sharing work from a math activity were as much a part of the classroom culture as sitting together to listen to a story, to talk about a new activity, or to anticipate an upcoming event.

Early in the school year, whole-class discussions are best kept short and focused. For example, after exploring pattern blocks, students might simply share experiences with the new material in a discussion structured almost as list-making:

What did you notice about pattern blocks? Who can tell us something different?

With questions like these, lots of students can participate without one student taking a lot of time.

Later in the year, when students are sharing their strategies for solving problems, you can use questions that allow many students to participate at once by raising their hands. For example:

Luke just shared that he solved the problem by counting out one cube for every person in our classroom. Who else solved the problem the same way Luke did?

In this way, you acknowledge the work of many students without everyone sharing individually.

Sometimes all students should have a chance to share their math work. You might set up a special "sharing shelf" or display area to set out or post student work. By gathering the class around the shelf or display, you can easily discuss the work of every student.

The ability to reflect on one's own thinking and to consider the ideas of others evolves over time, but even young students can begin to understand that an important part of doing mathematics is being able to explain your ideas and give reasons for your answers. In the process, they see that there can be many ways of finding solutions to the same problem. Over the year, your students will become more comfortable thinking about their solution methods, explaining them to others, and listening to their classmates explain theirs.

Games: The Importance of Playing More Than Once

> Teacher Note

Games are used throughout the *Investigations* curriculum as a vehicle for engaging students in important mathematical ideas. The game format is one that most students enjoy, so the potential for repeated experiences with a concept or skill is great. Because most games involve at least one other player, students are likely to learn strategies from each other whether they are playing cooperatively or competitively.

The more times students play a mathematical game, the more opportunities they have to practice important skills and to think and reason mathematically. The first time or two that students play, they focus on learning the rules. Once they have mastered the rules, their real work with the mathematical content begins.

For example, when students play the card game Compare, they practice counting and comparing two quantities up to 10. As they continue to play over days and weeks, they become familiar with the numerals to 10 and the quantities they represent. Later in the year, they build on this knowledge as they play Double Compare, a similar game in which they add and compare quantities up to 12. For many students, repeated experience with these two games leads them quite naturally to reasoning about numbers and number combinations, and to exploring relationships among number combinations.

Similarly, a number of games in *Pattern Trains and Hopscotch Paths* build and reinforce students' experience with repeating patterns. As students play Make a Train, Break the Train, and Add On, they construct and extend a variety of repeating patterns and are led to consider the idea that these linear patterns are constructed of units that repeat over and over again.

Games in the geometry unit *Making Shapes and Building Blocks,* such as Geoblock Match-Up, Build a Block, and Fill the Hexagons, expose students again and again to the structure of shapes and ways that shapes can be combined to make other shapes.

Students need many opportunities to play mathematical games, not just during math time, but other times as well: in the early morning as students arrive, during indoor recess, or as choices when other work is finished. Games played as homework can be a wonderful way of communicating with parents. Do not feel limited to those times when games are specifically suggested as homework in the curriculum; some teachers send home games even more frequently. One teacher made up "game packs" for loan, placing directions and needed materials in resealable plastic bags, and used these as homework assignments throughout the year. Students often checked out game packs to take home, even on days when homework was not assigned.

About Classroom Routines

Attendance

Taking the daily attendance and talking about who is and who is not in school are familiar activities in many kindergarten classrooms. Through the Attendance routine, students get repeated practice in counting a quantity that is significant to them: the number of people in their class. This is real data that they see, work with, and relate to every day. As they count the boys and girls in their class or the cubes in the attendance stick, they are counting quantities into the 20s. They begin to see the need to develop strategies for counting, including ways to double-check and to organize or keep track of a count.

Counting is an important mathematical idea in the kindergarten curriculum. As students count, they are learning how our number system is constructed, and they are building the knowledge they need to begin to solve numerical problems. They are also developing critical understandings about how numbers are related to each other and how the counting sequence is related to the quantities they are counting.

In *Investigations,* students are introduced to the Attendance routine during the first unit of the kindergarten sequence, *Mathematical Thinking in Kindergarten.* The basic activity is described here, followed by suggested variations for daily use throughout the school year.

The Attendance routine, with its many variations, is a powerful activity for 5- and 6-year-olds and one they never seem to tire of, perhaps because it deals with a topic that is of high interest: themselves and their classmates!

Materials and Preparation

The Attendance routine involves an attendance stick and name cards or "name pins" to be used with a display board. (Many teachers begin the year with name cards and later substitute name pins as a tool for recording the data.)

To make the attendance stick you need interlocking cubes of a single color, one for each class member, and dot stickers to number the cubes.

To make name cards, print each student's first name on a small card (about 2 by 3 inches). Add a photo if possible. If you don't have school photos or camera and film, you might ask students to bring in small photos of themselves from home.

For "name pins," print each student's name on both sides of a clothespin, being sure the name is right side up whether the clip is to the right or to the left.

Name cards might be displayed in two rows on the floor or on a display board. The board should have "Here" and "Not Here" sections, each divided into as many rows or columns as there are students in your class. To display name cards on the board, you might use pockets, cup hooks, or small pieces of Velcro or magnetic tape. Name pins can be clipped down the sides of a sturdy vertical board.

Collecting Attendance Data

How Many Are We? With the whole group, establish the total number of students in the class this year by going around the circle and counting the number of children present.

Encourage students to count aloud with you. The power of the group can often get the class as a whole much further in the counting sequence than many individuals could actually count. While one or two children may be able to count to the total number of students in the class, do not be surprised or concerned if, by the end of your count, you are the lone voice. Students learn the counting sequence and how to count by having many opportunities to count, and to see and hear others counting.

When you have counted those present, acknowledge any absent students and add them to the total number in your class.

About Classroom Routines

Counting Around the Circle Counting Around the Circle is a way to count and double-check the number of students in a group. Designate one person in the circle as the first person and begin counting off. That is, the first person says "1," the second person says "2," and so on around the circle. As students are learning how to count around the circle, you can help by pointing to the person whose turn it is to count. Some students will likely need help with identifying the next number in the counting sequence. Encouraging students to help each other figure out what number might come next establishes a climate of asking for and giving help to others.

Counting Around the Circle takes some time for students to grasp—both the procedure itself and its meaning. For some students, it will not be apparent that the number they say stands for the number of people who have counted thus far. A common response from kindergartners first learning to count off is to relate the number they say to a very familiar number, their age. Expect someone to say, for example, "I'm not 8, I'm 5!" Be prepared to explain that the purpose of counting off is to find out how many students are in the circle, and that the number 8 stands for the people who have been counted so far.

Representing Attendance Data

The Attendance Stick An attendance stick is a concrete model, made from interlocking cubes, that represents the total number of students in the classroom. For young students, part of knowing that there are 25 students in the class is seeing a representation of 25 students. The purpose of this classroom routine is not only to familiarize students with the counting sequence of numbers above 10, but also to help students relate these numbers to the quantities that they represent.

To make an attendance stick, distribute an interlocking cube to each student in the class. After counting the number of students present, turn their attention to the cubes.

We just figured out that there are [25] students in our classroom today. When you came to group meeting this morning, I gave everybody one cube. Suppose we collected all the cubes and snapped them together. How many cubes do you think we would have?

Collect each student's cube and snap them together into a vertical tower or stick. Encourage students to count with you as you add on cubes. Also add cubes for any absent students.

Ayesha is not here today. Right now our stick has 24 cubes in it because there are 24 students in school today. If we add Ayesha's cube, how many cubes will be in our stick?

Using small dot stickers, number the cubes. Display the attendance stick prominently in the group meeting area and refer to it each time you take attendance.

By counting around, we found that 22 of you are here today. Let's count up to 22 on the attendance stick. Count with me: 1, 2, 3 . . . *[when you reach 22, snap off the remaining cubes].* **So this is how many students are not here—who wants to count them?**

In this way, every day the class sees the attendance stick divided into two parts to represent the students HERE and NOT HERE.

Name Cards or Name Pins Name cards or pins are another concrete way to represent the students. Whereas the attendance stick represents *how many students* are in the class, name cards or pins provide additional data about *who* these people are.

Once students can recognize their name in print, they can simply select their card or pin from the class collection as they enter the classroom each day. At a group meeting, the names can be displayed to show who is here and who is not here, perhaps as a graph on the floor or on some type of display board.

Examining Attendance Data

Comparing Groups In addition to counting, the Attendance routine offers experience with part-whole relationships as students divide the total number into groups, such as PRESENT and ABSENT (HERE and NOT HERE) or GIRLS and BOYS. As they compare these groups, they are beginning to analyze the data and compare quantities: Which is more? Which is less? *How many* more or less? While the numbers for the groups can change on any given day, the sum of the two groups remains the same. Understanding part-whole relationships is a central part of both sound number sense and a facility with numbers.

The attendance stick and the name cards or name pins are useful tools for representing and comparing groups. One day you might use the attendance stick to count and compare how many students are present and absent; another day you might use name cards or pins the same way. Once students are familiar with the routine, you can represent the same data using more than one tool.

To compare groups, choose a day when everyone is in school. Count the number of boys and the number of girls.

Are there more boys than girls? How do you know? How many more?

Have the boys make a line and the girls make a line opposite them. Count the number of students in each line and compare the two lines.

Which has more? How many more?

Use the name cards or the attendance stick to double-check this information.

Once the total number of boys and girls is established, you can use this information to make daily comparisons.

Count the number of girls. Are all the girls HERE today? If not, how many are NOT HERE? How do you know? Can we show this information using the name cards? *[Repeat for the boys.]*

If we know that two girls and two boys are NOT HERE, how many in all are NOT HERE in school today? How do you know? Let's use the name cards to double-check.

When students are very familiar with this routine, with the total number in their class, and with making and comparing groups, you can pose a more difficult problem. For example:

If we know four students are NOT HERE in school today, how many students are HERE today? What are all the ways we can figure that out, without counting off?

Some students might suggest breaking four cubes off the attendance stick and counting the rest. Others might suggest counting back from the total number of students. Still others might suggest counting up from 4 to the total number of students.

In addition to being real data that students can see and relate to every day, attendance offers manageable numbers to work with. Repetition of this routine over the school year is important; only after students are familiar with the routine will they begin to focus on the numbers involved. Gradually, they will start to make some important connections between counting and comparing quantities.

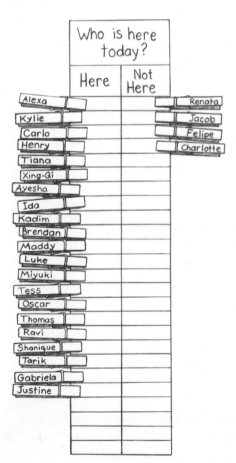

Appendix: About Classroom Routines ■ **105**

About Classroom Routines

Counting Jar

Counting is the foundation for much of the number work that students do in kindergarten and in the primary grades. Children learn to count by counting and hearing others count. Similarly, they learn about quantity through repeated experiences with organizing and counting sets of objects. The Counting Jar routine offers practice with all of these.

When students count sets of objects in the jar, they are practicing the counting sequence. As in the Attendance routine, they begin to see the need to develop strategies for counting, including ways to double-check and keep track of what they have counted. By recording the number of objects they have counted, students gain experience in representing quantity and conveying mathematical information to others. Creating a new equivalent set gives them not only another opportunity to count, but also a chance to compare the two amounts.

> Does my set have the same number as the set in the jar? How do I know?
>
> The jar has 8 and I have 7. I need 1 more because 8 is 1 more than 7.

As students work, they are developing a real sense of both numbers and quantities.

The Counting Jar routine is introduced in the first unit of the kindergarten curriculum, *Mathematical Thinking in Kindergarten*. The basic activity is described here, followed by suggested variations for use throughout the school year on a weekly basis.

Materials and Preparation

Obtain a clear plastic container, at least 6 inches tall and 4–5 inches in diameter. Fill it with a number of interesting objects that are uniform in size and not too small, such as golf or table tennis balls, small blocks or tiles, plastic animals, or walnuts in the shell. The total should be a number that is manageable for most students in your class; initially, 5 to 12 objects would be appropriate quantities.

Prepare a recording sheet on chart paper. At the top, write *The Counting Jar*, followed by the name of the material inside. Along the bottom, write a number line. Some students might use this number line to help them count objects or as a reference for writing numerals. Place each number in a box to clearly distinguish one from another.

About Classroom Routines

Laminate this chart so that students can record their counts on the chart with stick-on notes, write-on/wipe-off markers, or small scraps of paper and tape; these can later be removed and the chart reused.

Also make available one paper plate for each student and sets of countable materials, such as cubes, buttons, keys, teddy bear counters, or color tiles, so that students can create a new set of materials that corresponds to the quantity in the Counting Jar.

Counting

How Many in the Jar? This routine has three basic steps:

- Working individually or in pairs, students count the objects in the Counting Jar.
- Students make a representation that shows how many objects are in the jar and place their response on the chart.
- Students count out another set of objects equivalent to the quantity in the Counting Jar. They place this new set on a paper plate, write their name on the plate, and display their equivalent collection near the Counting Jar.

As you use the Counting Jar throughout the school year, call attention to it in a whole-group meeting whenever you have changed the material or the amount inside the jar. Then leave it in a convenient location for two or three days so that everyone has a chance to count. After most students have counted individually, meet with the whole class and count the contents together.

Note: Some kindergarten teachers use a very similar activity for estimation practice. We exclude the task of estimation from the basic activity because until students have a sense of quantity, a sense of how much 6 is, a sense of what 10 balls look like compared to 10 cubes, it is difficult for them to estimate or predict how large a quantity is. When students are more familiar with the routine and have begun to develop a sense of quantity, you might include the variations suggested for estimation.

One More, One Less When students can count the materials in the jar with a certain amount of accuracy and understanding, try this variation for work with the ideas "one more than" and "one less than." As you offer the Counting Jar activity, ask students to create a set of objects with one more (or less) than the amount in the jar.

Filling the Jar Ourselves When the Counting Jar routine is firmly established, give individuals or pairs of students the responsibility for filling the jar. Discuss with them an appropriate quantity to put in the jar or suggest a target number, and let students decide on suitable objects to put in the jar.

At-Home Counting Jars Suggest to families that they set up a Counting Jar at home. Offer suggestions for different materials and appropriate quantities. Family members can take turns putting sets of objects in the jar for others to count.

Estimation

Is It More Than 5? To introduce the idea of estimation, show students a set of five objects identical to those in the Counting Jar. This gives students a concrete amount for reference to base their estimate on. As they look at the known quantity, ask them to think about whether there are *more than* five objects in the jar. The number in the reference group can grow as the number of objects in the jar changes, and you can begin to ask "Is the amount in the jar more than 8? more than 10?"

More or Less Than Yesterday? You can also encourage students to develop estimation skills when the material in the jar stays the same over several days but the quantity changes. In this situation, students can use reasoning like this:

> Last time, when there were 8 blocks in the jar, it was filled up to *here*. Now it's a little higher, so I think there are 10 or 11 blocks.

About Classroom Routines

Calendar

"Calendar," with its many rituals and routines, is a familiar kindergarten activity. Perhaps the most important idea, particularly for young students, is viewing the calendar as a real-world tool that we use to keep track of time and events. As students work with the calendar, they become more familiar with the sequence of days, weeks, and months, and relationships among these periods of time. Time and the passage of time are challenging ideas for most 5- and 6-year-olds, and the ideas need to be linked to their own direct experiences. For example, explaining that an event will occur *after* a child's birthday or *before* a familiar holiday will help place that event in time for them.

The Calendar routine is introduced in the first unit of the kindergarten curriculum, *Mathematical Thinking in Kindergarten.* The basic activity is described here, followed by suggested variations for daily use throughout the school year.

Materials and Preparation

In most kindergarten classrooms, a monthly calendar is displayed where everyone can see it when the class gathers as a whole group. A calendar with date cards that can be removed or rearranged allows for greater flexibility than one without. Teachers make different choices about how to display numbers on this calendar. We recommend displaying all the days, from 1 to 30 or 31, all month long. This way the sequence of numbers and the total number of days are always visible, thus giving students a sense of the month as a whole.

You can use stick-on labels to highlight special days such as birthdays, class trips or events, non-school days, or holidays. Similarly, find some way to identify *today* on the calendar. Some teachers have a special star or symbol to clip on today's date card, or a special tag, much like a picture frame, that hangs over today's date.

A Sense of Time

The Monthly Calendar When first introducing the calendar, ask students what they notice. They are likely to mention a wide variety of things, including the colors they see on the calendar, pictures, numbers, words, how the calendar is arranged, and any special events they know are in that particular month. If no one brings it up, ask students what calendars are for and how we use them.

At the beginning of each month, involve students in organizing the dates and recording special events on the calendar. The following questions help them understand the calendar as a tool for keeping track of events in time:

> If our trip to the zoo is on the 13th, on which day should we hang the picture of a lion?
>
> Is our trip tomorrow? the next day? this week?
>
> What day of the week will we go to the zoo?

How Much Longer? Many students eagerly anticipate upcoming events or special days. Ask students to figure out how much longer it is until something, or how many days have passed since something happened. For example:

> How many more days is it until Alexa's birthday?
>
> Today is November 4. How many more days is it until November 10?
>
> How many days until the end of the month?
>
> How many days have gone by since our parent breakfast?

Ask students to share their strategies for finding the number of days. Initially many students will

count each subsequent day. Later some students may begin to find answers by using their growing knowledge of calendar structure and number relationships:

> I knew there were three more days in this row, and I added them to the three in the next row. That's six more days.

Calculating "how many more days" on the calendar is not an easy task. Quite likely students will not agree on what days to count. Consider the following three good answers, all different, to this teacher's question:

Today is October 4. Ida's birthday is on October 8. How many more days until her birthday?

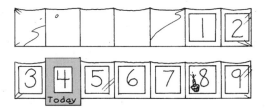

Tess: I think there are four more days because it's 4 . . . *[counting on her fingers]* 5, 6, 7, 8.

Ravi: There are three more days. See? *[He points to the three calendar dates between October 4 and October 8—5, 6, and 7—and counts three date cards.]*

Gabriela: It's five more days until her birthday. *[Using the calendar, she points to today and counts "1, 2, 3, 4, 5," ending on October 8.]*

All of these students made sense of their answers and, considering their reasoning, all three were correct. That's why, when asking "how many more?" questions based on the calendar, it is important also to ask students to explain their thinking.

Numbers on the Calendar

Counting Days The calendar is a place where students can daily visit and become more familiar with the sequence of counting numbers up to 31. Because the numbers on the calendar represent the number of days in a month, the calendar is actually a way of *counting days*. You can help students with this idea:

Today is September 13. Thirteen days have already gone by in this month. If we start counting on 1, what number do you think we will end up on? Let's try it.

As you involve students in this way, they have another chance to see that numbers represent a quantity, in this case a number of days.

Missing Numbers or Mixed-Up Numbers Once students are familiar with the structure of the calendar and the sequence of numbers, you can play two games that involve removing and rearranging the dates. To play Missing Numbers, choose two or three dates on the monthly calendar and either remove or cover them. As students guess which numbers are missing, encourage them to explain their thinking and reasoning. Do they count from the number 1 or do they count on from another number? Do they know that 13 comes *after* 12 and *before* 14?

Mixed-Up Numbers is played by changing the position of numbers on the calendar so that some are out of order. Students then fix the calendar by pointing out which numbers are out of order.

Patterns on the Calendar

Looking for Patterns Some teachers like to point out patterns on the calendar. The repeating sequence of the days of the week and the months of the year are patterns that help students explore the cyclical nature of time. Many students quickly recognize the sequence of numbers 1 to 30 or 31, and some even recognize another important pattern on the calendar: that the columns increase by 7. However, in order to maintain the focus on the calendar as a tool for keeping track of time, we recommend using the Calendar routine only to note patterns that exist within the structure of the calendar and the sequence of days and numbers. The familiar activity of adding pictures or shapes to form repeating patterns can be better done in another routine, Patterns on the Pocket Chart.

About Classroom Routines

Today's Question

Collecting, representing, and interpreting information are ongoing activities in our daily lives. In today's world, organizing and interpreting data are vital to understanding events and making decisions based on this understanding. Because young students are natural collectors of materials and information, working with data builds on their natural curiosity about the world and people.

Today's Question offers students regular opportunities to collect information, record it on a class chart, and then discuss what it means. While engaged in this data collection and analysis, students are also counting real, meaningful quantities (How many of us have a pet?) and comparing quantities that are significant to them (Are there more girls in our class or more boys?). When working with questions that have only two responses, students explore part-whole relationships as they consider the total number of answers from the class and how that amount is broken into two parts.

Today's Question is introduced in the first unit of the kindergarten curriculum, *Mathematical Thinking in Kindergarten*. The basic routine is described here, followed by variations. Plan to use this routine throughout the school year on a weekly basis, or whenever a suitable and interesting question arises in your classroom.

Materials and Preparation

Prepare a chart for collecting students' responses to Today's Question. If you plan to use this routine frequently, either laminate a chart so that students can respond with wipe-off markers, or set up a blank chart on 11-by-17-inch paper and make multiple photocopies. The drawback of a laminated wipe-off chart is that you cannot save the information collected; with multiple charts, you can look back at data you have collected earlier or compare data from previous questions.

Make a section across the top of the chart, large enough to write the words *Today's Question* followed by the actual question being asked.

Mark the rest of the chart into two equal columns (later, you may want three columns). Leave enough space at the top of each column for the response choices, including words and possibly a sketch as a visual reminder.

Leave the bottom section (the largest part of the chart) blank for students to write their names to indicate their response. Your chart will look something like this:

Later in the year, you may want a chart with write-on lines in the bottom section to help students to compare numbers of responses in the two or three categories. Be sure to allow one line for each child in the class. Lines are also helpful guides if you collect data with "name pins," or clothespins marked on both sides with student names, as suggested for the Attendance routine (see illustration on p. 105).

Choosing Questions

Especially during the first half of the school year, try to choose questions with only two responses. With two categories of data, students are more likely to see the part-whole relationship between the number of responses in each category and the total number of students in the class.

About Classroom Routines

As your students become familiar with the routine and with analyzing the data they collect, you may decide to add a third response category. This is useful for questions that might not always elicit a clear yes-or-no response, such as these:

Do you think it will rain? *(yes, no, maybe)*

Do you want to play outside today? *(yes, no, I'm not sure)*

Do you eat lunch at school? *(yes, no, sometimes)*

As you choose questions and set up the charts for this routine, consider the full range of responses and modify or drop the question if there seem to be too many possible answers. Later in the year, as students become familiar with this routine, you may want to involve them in organizing and choosing Today's Question.

Questions About the Class With Today's Question, students can collect information about a group of people and learn more about their classmates. For example:

Are you a boy or a girl?

Are you 5 or 6 years old?

Do you have a younger brother?

Do you have a pet?

Did you bring your lunch to school today?

Do you go to an after-school program?

Do you like ice cream?

Did you walk or ride to school this morning?

Some teachers avoid questions about potentially sensitive issues (Have you lost a tooth? Can you tie your shoes?), while others use this routine to carefully raise some of these issues. Whichever you decide, it is best to avoid questions about material possessions (Does your family have a computer?).

Questions for Daily Decisions When you pose questions that involve students in making decisions about their classroom, they begin to see that they are collecting real data for a purpose. These data collection experiences underscore one of the main reasons for collecting data in the real world: to help people make decisions. For example:

Which book would you like me to read at story time? (Display two books.)

Would you prefer apples or grapes for snack?

Should we play on the playground or walk to the park today?

Questions for Curriculum Planning Some teachers use this routine to gather information that helps them plan the direction of a new curriculum topic or lesson. For example, you can learn about students' previous experiences and better prepare them before reading a particular story, meeting a special visitor, or going on a field trip, with questions like these:

Have you ever read or heard this story?

Have you ever been to the science museum?

Have you ever heard of George Washington?

For questions of this type, you might want to add a third possible response (*I'm not sure* or *I don't know*).

Discussing the Data

Data collection does not end with the creation of a representation or graph to show everyone's responses. In fact, much of the real work in data analysis begins after the data has been organized and represented. Each time students respond to Today's Question, it is important to discuss the results. Consider the following questions to promote data analysis in classroom discussions:

What do you think this graph is about?

What do you notice about this graph?

What can you tell about [the favorite part of our lunch] by looking at this graph?

If we went to another classroom, collected this same information, and made a graph, do you think that graph would look the same as or different from ours?

Graphs and other visual representations of the data are vehicles for communication. Thinking about what a graph represents or what it is communicating is a part of data analysis that even the youngest students can and should be doing.

About Classroom Routines

Patterns on the Pocket Chart

Mathematics is sometimes called "the science of patterns." We often use the language of mathematics to describe and predict numerical or geometrical regularities. When young students examine patterns, they look for relationships among the pattern elements and explore how that information can be used to predict what comes next. The classroom routine Patterns on the Pocket Chart offers students repeated opportunities to describe, copy, extend, create, and make predictions about repeating patterns. The use of a 10-by-10 pocket chart to investigate patterns of color and shape builds a foundation for the later grades, when this same pocket chart will display the numbers 1 to 100 and students will investigate patterns in the arrangement of numbers.

This routine is introduced in the second unit of the kindergarten curriculum, *Pattern Trains and Hopscotch Paths*. The basic routine is described here, followed by variations for use throughout the school year on a weekly basis.

Materials and Preparation

For this routine you will need a pocket chart, such as the vinyl Hundred Number Wall Chart (with transparent pockets and removable number cards). You will also need 2-inch squares of construction paper of different colors, a set of color tiles (ideally, the colors of the paper squares will match the tiles), and a set of 20–30 What Comes Next? cards. These cards, with a large question mark in the center, are cut slightly larger than 2 inches so they will cover the colored squares. A blackline master for these cards is provided in the unit *Pattern Trains and Hopscotch Paths*. You can easily make your own cards with tagboard and a marking pen.

For the variation Shapes, Shells, and Such, you can use math manipulatives such as pattern blocks and interlocking cubes, picture or shape cards, or collections of small objects, such as buttons, keys, or shells. The only limitation is the size of the pockets on your chart.

What Comes Next?

Before introducing this activity, arrange an a-b repeating pattern in the first row of the pocket chart using ten paper squares in two colors of your choice. Beginning with the fifth position, cover each colored square with a What Comes Next? (question mark) card.

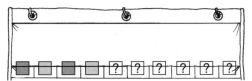

Gather students where the pocket chart is clearly visible and they have a place to work with color tiles, either on the floor or at tables.

Begin by asking students what they notice about the chart. Some may comment on the structure of the chart, some on the two-color pattern, and others may notice the question marks. Explain that each time they see one of these question marks, they should think "What comes next?" and decide what color might be under that card.

Provide each pair with a small cup of color tiles that match the paper squares. Ask students to build the first part of the pattern with color tiles and then predict what color comes next.

Who can predict, or guess, what color is hidden under each question mark on our chart? Use the tiles in your cup to show me what color would come next. How do you know?

Now, with your partner, see if you can make this pattern longer, using the tiles in your cup. Stop when your pattern has ten tiles.

When everyone has made a longer pattern, "read" the pattern together as a whole class. Verbalizing the pattern they are considering often helps students internalize it, recognize any errors in the pattern, and determine what comes next.

This basic activity can be done quickly, especially if students do not build the pattern with tiles. Many teachers integrate this routine into their

About Classroom Routines

group meeting time on a regular basis, making one or two patterns on the pocket chart and asking students to predict what comes next.

Initially, use only two colors or two variables in the patterns. In addition to a-b (for example, blue-green) repeating patterns, build two-color patterns such as a-a-b (blue-blue-green), a-b-b (blue-green-green), or a-a-b-b (blue-blue-green-green).

Variations

Making Longer Patterns When students are familiar with the basic activity, they can investigate what happens to an a-b pattern when it "wraps around" and continues to the next line. If the pattern continues in a left-to-right progression, the pattern that emerges is the same one older students see when they investigate the patterns of odd and even numbers on the 100 chart.

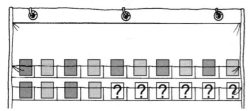

Shapes, Shells, and Such Color is just one variable for patterns; others can be made using a wide variety of materials and pictures.

shell, shell, button, shell, shell, button

triangle, square, triangle, square

♡ ☾ ♡ ☾ ♡ ☾

⇧ ⇩ ⇩ ⇧ ⇩ ⇩ ⇧ ⇩

Picture cards, sometimes used by kindergarten teachers to make patterns on the calendar, make great patterns on the pocket chart without the distraction of the calendar elements.

What Comes *Here*? Predicting what comes next is an important idea in learning about patterns. Also important is being able to look ahead and predict what comes *here?* even further down the line. Instead of asking for the *next* color in a pattern sequence, point to a pocket three or four squares along and ask students to predict the color under that question mark. As you collect responses, ask students to explain how they predicted that color.

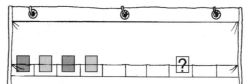

Border Patterns Explore a repeating pattern that extends around the entire outer edge of the pocket chart. Begin by filling the top row of the chart and asking what color would come next if this pattern turned the corner and went down the right side of the chart. Continue adding squares to finish the border. Every few days, begin a new pattern and ask students to help you complete the border. Start with a-b patterns. Gradually vary the pattern type, but continue to use only two colors, trying patterns such as a-a-b, a-b-b, a-a-b-b, or a-a-b-a. Ask students to notice which types make a continuous pattern all around the border and which do not.

With any border pattern, you can include a few What Comes Next? cards and ask students to predict the color of a particular pocket.

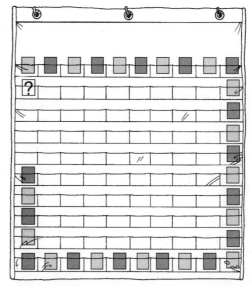

Patterns for Choice Time Hang the pocket chart where students can reach it. During free time or Choice Time, two or three students can work together to make their own pattern on the pocket chart, using colored paper squares or color tiles.

TIPS FOR THE LINGUISTICALLY DIVERSE CLASSROOM

It is likely that more students with limited English proficiency will be enrolled in kindergarten than any other grade. Moreover, many will be at the earliest stages of language acquisition. By correctly identifying a student's current level of English, you can create appropriate stimuli to ensure successful communication when presenting activities from *Investigations*.

The four stages of language acquisition are characterized as follows:

- **Preproduction** Students do not speak the language at this stage; they are dependent upon modeling, visual aids, and context clues to obtain meaning.
- **Early production** Students begin to produce isolated words in response to comprehensible questions. Responses are usually yes, no, or other single-word answers.
- **Speech emergence** Students at this level now have a limited vocabulary and can respond in short phrases or sentences. Grammatical errors are common.
- **Intermediate fluency** Students can engage in conversation, producing full sentences.

You need to be aware of these four levels of proficiency while applying the following tips. The goal is always to ensure that students with limited English proficiency develop the same understandings as their English-speaking peers as they participate in this unit.

Tips for Small-Group Work Whenever possible, pair students with the same linguistic background and encourage them to complete the task in their native language. Students are more likely to have a successful exchange of ideas when they speak the same language. In other situations, teach all students how to make their communications comprehensible. For example, encourage students to point to objects they are discussing.

Tips for Whole-Class Activities To keep whole-group discussions comprehensible, draw simple sketches or diagrams on the board to illustrate key words; point to objects being discussed; use contrasting examples to help explain the attribute under discussion; model all directions; choose students to model activities or act out scenarios.

Tips for Observing the Students Assessment in the kindergarten units is based on your observations of students as they work, either independently or in groups. At times you will intervene by asking questions to help you evaluate a student's understanding. When questioning students, it is crucial not to misinterpret responses that are incomplete simply because of linguistic difficulties.

In many cases, students may understand the mathematical concept being asked about but not be able to articulate their thoughts in English. You need to formulate questions that allow students to respond at their stage of language acquisition in a way that indicates their mathematical understanding.

For example, the following question, focusing on students' thinking about the attributes of three-dimensional shapes, is appropriate for students at the speech emergence and intermediate fluency stages of language acquisition: "How do students describe the different Geoblocks?"

The question could be reworded as follows for students at the preproduction stage to elicit nonverbal responses: "Show me a block that is smaller than this one *[point to a block]*. Can you show me a block with a triangle *[outline a triangle shape in the air with your finger]* on it somewhere?"

For students at the early production stage, use a question like the following to ask for a one-word response: "Is the face of this block *[outline the face with your finger]* a square?"

As you observe the students working, keep in mind which guidelines are appropriate for students at the different stages of language acquisition. Following is a categorization of typical questions from this unit.

Questions appropriate for students at the preproduction stage:

- Do students create a pattern around the border of their page [for the Book of Shapes]? What kind of pattern? Does the pattern include shapes?
- Do they have a sense that pairs or combinations of [pattern] blocks can be substituted for other blocks? For example, do they substitute two trapezoids for a hexagon, or two triangles for a blue rhombus?
- [When working on the computer in *Shapes*], are students able to select blocks and move them across the screen? Are they using the Turn tool to turn or rotate pattern blocks?
- Are students able to make a variety of [clay] shapes? Do they consult the shape posters, or do they seem to "just know" the shapes they want to construct?
- Do students randomly choose smaller [Geoblocks] and try to build a larger one [for Build a Block]? Do they seem to recognize and specifically select blocks that are half of a larger block and put two of them together? Do they use smaller cubes to build the larger cubes?
- Do students keep in mind all the faces on a block as they work [on Matching Faces]? That is, when a particular face doesn't match one face of another block, do they check to see if it matches any of the other faces on that block?

Questions appropriate for students at the early-production and early-speech-emergence stages:

- How do students talk about and describe shapes [while on the Shape Hunt]? Do they use a 2-D name to refer to a 3-D shape? For example, do they say "circle" for a sphere or "square" for a cube?
- Are students able to make a shape using the clay? How do they describe their shape? Do they refer to the shape by name?
- What vocabulary do students use to describe squares, rectangles, and triangles [in Geoblock Match-Up]? Does their language describe differences between the shapes? For example, do they talk about shapes that are *thick* and *thin*, or *tall* and *short*?

Questions appropriate for students at the late-speech-emergence and intermediate-fluency stages:

- Are students able to describe the shape of an object [they find on the Shape Hunt]? Can they tell in what ways it is the same as one of the geometric solids?
- What attributes of shape do students notice [when talking about their pictures for the Book of Shapes]? Do they mention the number of sides? relationship between those sides? size of the shape? orientation of the shape? What language do they use to describe these attributes?
- How do students talk about and describe the pattern block shapes [while playing Fill the Hexagons]? How do they talk about and describe the spaces that remain in a partially-filled hexagon? (For example, in a hexagon outline with one trapezoid block already placed, students might hope for a roll of "the red one," "the half one," or "the trapezoid.")

VOCABULARY SUPPORT FOR SECOND-LANGUAGE LEARNERS

The following activities will help ensure that this unit is comprehensible to students who are acquiring English as a second language. The suggested approach is based on *The Natural Approach: Language Acquisition in the Classroom* by Stephen D. Krashen and Tracy D. Terrell (Alemany Press, 1983). The intent is for second-language learners to acquire new vocabulary in an active, meaningful context.

Note that *acquiring* a word is different from *learning* a word. Depending on their level of proficiency, students may be able to comprehend a word upon hearing it during an investigation, without being able to say it. Other students may be able to use the word orally, but not read or write it. The goal is to help students naturally acquire targeted vocabulary at their present level of proficiency.

outline, fill in

1. Draw heavy outlines of shapes on blank index cards, or use copies of the Make-a-Shape Cards (pp. 178–180). Divide the cards into three arbitrary groups.

2. Run your finger around the shape on a card, explaining that this is the *outline* of a shape. Repeat with several cards.

3. Model using a crayon to *fill in* each outline in one group. Then model filling in part of the shape in another group. Finally, show that you will *not* fill in the shapes in the remaining group.

4. Sketch and identify three outlines on the board, with one outline all filled in, one partly filled in, and one not filled in.

5. Turn over all the cards in a stack and mix them. Tell students that this is a guessing game, and they are to guess which category the top card will fall in. Point to the top card in the stack and to the three categories on the board, asking for guesses. Students can guess by pointing to the category on the board (which you can then verbalize), or by naming the category.

6. After everyone has guessed, turn over the card to reveal the outline. Continue through the entire stack.

SHAPES

Teacher Tutorial

Contents

Overview	119
Getting Started with *Shapes*	
About *Shapes*	120
What Should I Read First?	120
Starting *Shapes*	120
How to Start an Activity	121
Free Explore	
About Free Explore	122
Building a Picture	122
More About Building Pictures	127
Using *Shapes*	
Tool Bar, Shape Bar, and Work Window	137
Arrow Tool: Dragging and Selecting	139
Turn, Flip, and Magnifying Tools: Moving and Sizing	140
Glue and Hammer Tools: Combining and Breaking Apart	141
Duplicate Tool and Pattern Button: Making Repeating Patterns	143
Freeze and Unfreeze Tools: Combining and Breaking Apart	144
Using Menu Commands	144
***Shapes* Menus**	
About Menus	145
Trouble-Shooting	
No *Shapes* Icon to Open	148
Nothing Happened After Double-Clicking on the *Shapes* Icon	148
In the Wrong Activity	148
A Window Closed by Mistake	148
Windows or Tools Dragged to a Different Position by Mistake	149
I Clicked Somewhere and Now *Shapes* Is Gone! What Happened?	149
How Do I Select a Section of Text?	149
System Error Message	149

I Tried to Print and Nothing Happened	149
I Printed the Work Window but Not Everything Printed	149

***Shapes* Messages** — 150

How to Install *Shapes* on Your Computer — 151
    Optional — 152

Other Disk Functions
    Saving Work on a Different Disk — 153
    Deleting Copies of Student Work — 153

Overview

The units in *Investigations in Number, Data, and Space®* ask teachers to think in new ways about mathematics and how students best learn math. Units such as *Making Shapes and Building Blocks* add another challenge for teachers—to think about how computers might support and enhance mathematical learning. Before you can think about how computers might support learning in your classroom, you need to know what the computer component is, how it works, and how it is designed to be used in the unit.

The *Shapes* Tutorial is written for you as an adult learner, as a mathematical explorer, as an educational researcher, as a curriculum designer, and finally—putting all these together—as a classroom teacher. It is not intended as a walk-through of the student activities in the unit. Rather, it is meant to provide experience using the computer program *Shapes* and to familiarize you with some of the mathematical thinking in the unit.

The first part of the Tutorial shows how to create a picture using the pattern block shapes. Through making this picture, you become familiar with the tools available in the *Shapes* software. The later parts of the Tutorial include more detail about each component of *Shapes* and can be used for reference. There is also detailed help available in the *Shapes* program itself.

Teachers new to using computers and *Shapes* can follow the detailed step-by-step instructions. Those with more experience might not need to read each step. As is true with learning any new approach or tool, you will test out hypotheses, make mistakes, be temporarily stumped, go down wrong paths, and so on. This is part of learning but may be doubly frustrating because you are dealing with computers. It might be helpful to work through the Tutorial and the unit in parallel with another teacher. If you get particularly frustrated, ask for help from the school computer coordinator or another teacher more familiar with using computers. It is not necessary to complete the Tutorial before beginning to teach the unit. You can work through in parts, as you prepare for parallel investigations in the unit.

Although the Tutorial will help prepare you for teaching the unit, you will learn most about *Shapes* and how it supports the unit as you work side by side with your students.

Getting Started with *Shapes*

About *Shapes*

Shapes is a computer manipulative, a software version of pattern blocks, that extends what students can do with these shapes. Students create as many copies of each shape as they want and use computer tools to move, combine, and duplicate these shapes to make pictures and designs and to solve problems. In addition to the standard pattern block shapes, the pattern block set in *Shapes* includes a quarter circle, which extends students' explorations to include shapes with curved sides.

What Should I Read First?

Read the next section, Starting *Shapes*, for specific information on how to load the *Shapes* program and choose an activity.

The section Free Explore takes you step by step through an example of working with *Shapes*. Read this section for a sense of what the program can do.

The section Using *Shapes* provides detailed information about each aspect of *Shapes*. Read this to learn *Shapes* thoroughly or to answer specific questions.

Starting *Shapes*

Note: These directions assume that *Shapes* has been installed on the hard drive of your computer. If not, see How to Install *Shapes* on Your Computer, p. 151.

 1. Turn on the computer by doing the following:

   a. If you are using an electrical power surge protector, switch to the ON position.

   b. Switch the computer (and the monitor, if separate) to the ON position.

   c. Wait until the desktop or workspace appears.

 2. Open *Shapes* by doing the following:

   a. Double-click on the *Shapes* Folder icon if it is not already open. To double-click, click twice in rapid succession without moving the pointer.

   b. Double-click on the *Shapes* icon in this folder.

120 ■ *Appendix: Shapes Teacher Tutorial*

c. Wait until the *Shapes* opening screen appears. Click on the bar "Click on this window to continue." when the message appears.

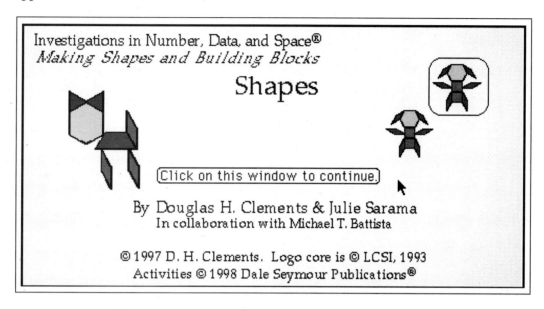

Start an activity by doing the following:

Click on Free Explore (or any activity you want).

How to Start an Activity

When you choose an activity, the Tool bar, Shape bar, and Work window fill the screen.

The following section provides a step-by-step example of working with *Shapes*.

Appendix: Shapes Teacher Tutorial ■ **121**

Free Explore

About Free Explore

The Free Explore activity is available for you to use as an environment to explore *Shapes*. It can also be used to extend and enhance activities.

When you choose Free Explore, you begin with an empty Work window. You can build a picture in that window with the shapes from the Shape bar. The tools in the Tool bar enable you to move, duplicate, and glue the shapes you select from the Work window.

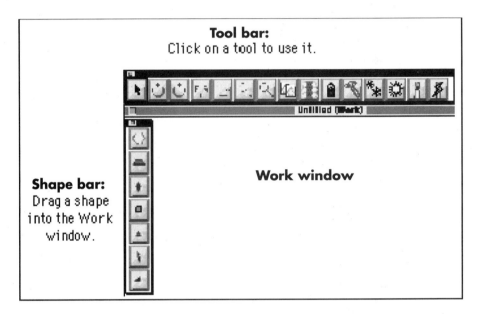

Building a Picture

Let's begin by making a building in the Work window.

☛ 1. Build the front of the building by doing the following:

a. Drag an orange square shape off the Shape bar.

| Move the cursor so it is on the square. It becomes a hand. | Click the mouse button and hold it down . . . | . . . while you move the square where you want it. |

b. Slide the square to the middle of the Work window.

If you need to move the square, just click on it and drag it again.

122 ■ *Appendix: Shapes Teacher Tutorial*

c. Drag another orange square shape from the Shape bar and place it next to the first one.

Notice that the new square "snaps" right next to the first one.

d. Continue this procedure until the "front" of the building is finished.

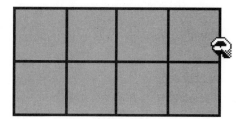

☛ 2. Build the side of the building by doing the following:

Drag two tan rhombuses (thin diamonds) for the side of the building from the Shape bar and slide into place.

Young students might not do this, but we're going to try for a three-dimensional effect!

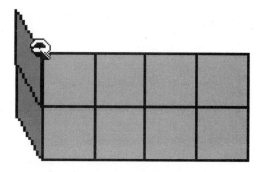

☛ 3. Start the roof of the building by doing the following:

Drag a blue rhombus (diamond) from the Shape bar.

This shape will have to be turned to make it fit.

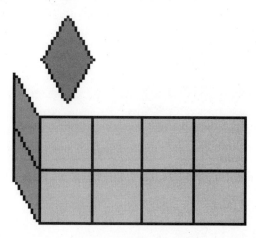

☞ 4. Turn the blue rhombus by doing the following:

   a. Click on the Left Turn tool in the Tool bar.

   The cursor changes from an arrow to a left-turn circle. ↺

   b. Click this new cursor on the blue rhombus.
   The rhombus turns to the left.

   c. Click on the blue rhombus a second time.
   The rhombus turns to the left again, and it is ready to be slid into place.

☞ 5. Slide the blue rhombus into place by doing the following:

   a. Click on the Arrow tool in the Tool bar.
   The cursor changes back to an arrow.

   b. Slide the rhombus into place.

   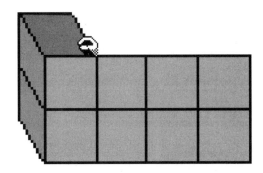

☛ 6. Duplicate the blue rhombus three times to make three copies of it:

   a. Click on the Duplicate tool in the Tool bar.

   The cursor changes from an arrow to the Duplicate icon .

   b. Click the Duplicate tool *on* the blue rhombus.
   A duplicate is made. Using the Duplicate tool is particularly appropriate in this case because the duplicate is turned the correct way automatically.

   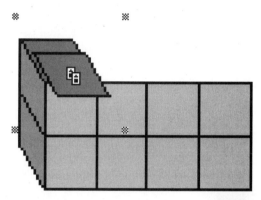

   c. Click on the Arrow tool in the Tool bar.
   The cursor changes back to an arrow.

   d. Slide the duplicate rhombus into place.

   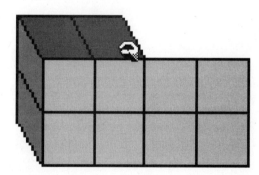

Appendix: Shapes Teacher Tutorial ■ **125**

e. Repeat steps a through d to make and position two more duplicates.

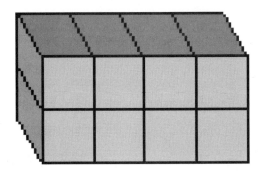

Note: To be more efficient, we could have duplicated three copies right away and then slid each into place one after the other.

☛ 7. Make a half-circle doorway.

   a. Drag a quarter circle from the Shape bar. Slide it into place.

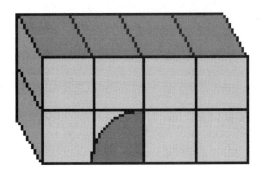

   b. Drag another quarter circle (or duplicate the first one) and slide it into place.

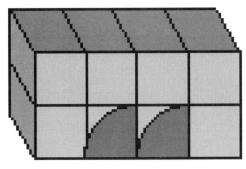

   c. Click on the Vertical Flip tool in the Tool bar.

   The cursor changes from an arrow to the Vertical Flip icon F⋮꜒.

126 ■ *Appendix: Shapes Teacher Tutorial*

d. Click the Vertical Flip tool on the second quarter circle.

The shape flips over a vertical line through the center of the shape. Because the quarter circle is symmetric, we could have also turned it several times to the right, but flipping is more efficient.

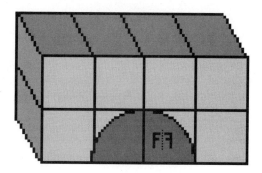

Our building is finished. Now let's add some more parts to the picture.

---

 1. Make a sun.

More About Building Pictures

a. Drag a yellow hexagon from the Shape bar.

Place it in the upper right-hand corner of the Work window.

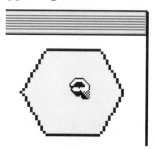

b. Get another yellow hexagon and place it right over the first one.

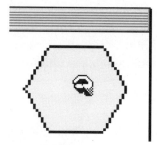

c. Click on the Right Turn tool in the Tool bar.

The cursor changes from an arrow to the Right Turn icon .

Appendix: Shapes Teacher Tutorial ■ **127**

d. Click the Right Turn tool on the second hexagon.

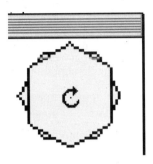

Our sun is finished. Now let's add a walkway. It will be a pattern of several shapes. First, we'll design the unit.

☞ 1. Make a unit for the walkway.

a. Drag a yellow hexagon, a red trapezoid, a green triangle, and a blue rhombus from the Shape bar.

Place the shapes in the lower left-hand corner of the Work window as shown. Some shapes will have to be turned to make the pattern shown.

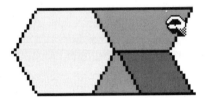

b. Click on the Glue tool in the Tool bar.

The cursor changes from an arrow to the Glue icon .

c. Click in the middle of *each* of the four shapes in the unit.

The cursor changes to a "squirt glue" icon whenever you click on a shape that has not yet been glued. Note that you have to click on each shape; even though they are "snapped" and touching sides, you must indicate each one you want glued together by clicking in the middle of each shape.

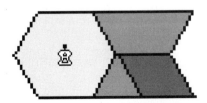

The four shapes are now a glued group. They can be moved, turned, flipped, or duplicated as if they were one shape.

128 ■ *Appendix:* Shapes *Teacher Tutorial*

You can check that the shapes are one glued group. Move the cursor to the Glue tool in the Tool bar and hold down the mouse button. The following will appear on your screen, indicating one group.

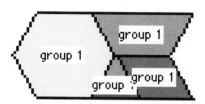

☛ 2. Duplicate the unit for the walkway.

   a. Click on the Duplicate tool in the Tool bar.

   The cursor changes from an arrow to the Duplicate icon .

   b. Click the Duplicate tool on the group of shapes.
   A duplicate is made. Using the Duplicate tool is necessary because we're going to make a repeating pattern.

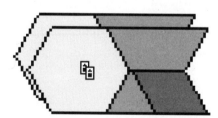

☛ 3. Define the motion for the pattern.

   a. Click on the Arrow tool in the Tool bar.
   The cursor changes back to an arrow.

   b. Slide the duplicate of the unit where you want it to be for the start of the pattern.

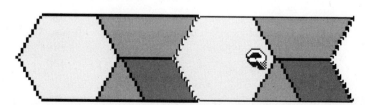

Appendix: Shapes Teacher Tutorial ■ **129**

☞ 4. Continue the pattern.

   a. Click on the Pattern button in the Tool bar.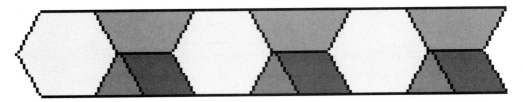
      The Pattern tool is a button. Simply clicking on a button will perform the action immediately. The next part of the pattern is put into place.

   b. Keep clicking on the Pattern button until your walkway extends across the window.

Our walkway is finished. Now let's make trees. Use the Pattern tool again to make the top of the first tree.

☞ 1. Make a unit for the treetop.

   Get a green triangle from the Shape bar.

   Place it on the left-hand side of the Work window. We don't need to glue this time because our unit is just one shape.

☞ 2. Duplicate it.

   a. Click on the Duplicate tool in the Tool bar.

   The cursor changes from an arrow to the Duplicate icon .

   b. Click the Duplicate tool on the triangle.

☞ 3. Define the motion for the pattern.

   a. Click on the Right Turn tool in the Tool bar.

   The cursor changes back to the Right Turn icon ↻.

   b. Turn the duplicate of the unit to the right two times.

130 ■ *Appendix:* Shapes *Teacher Tutorial*

c. Click on the Arrow tool in the Tool bar.
   The cursor changes back to an arrow.

d. Slide the duplicate of the unit where you want it to be for the start of the pattern.

   We have now defined the motion for the pattern.

4. Continue the pattern.

   a. Click on the Pattern button in the Tool bar four times to complete the treetop.

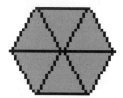

We'll finish the tree.

☞ 1. Make a trunk and ground cover.

   a. Drag two squares from the Shape bar. Place the squares below the green triangles as shown.

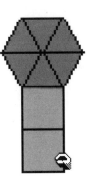

   b. Drag three tan rhombuses from the Shape bar. Turn them left three times and place them as ground cover. (You could also get one, turn it left three times, then use the Duplicate tool to make two copies.)

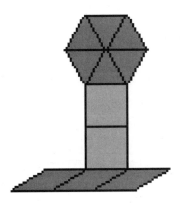

We need ground cover in back of the tree too.

   c. Drag three more tan rhombuses from the Shape bar. Place them in "back" of the others. (You could use the Duplicate tool to do this.)

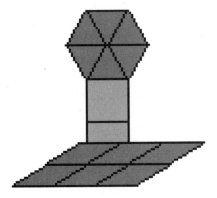

At this point the tree is behind the ground cover, but we can fix that.

☞ 2. Bring the tree trunk to the front of the ground cover.

   a. Select the bottom orange square by clicking in the middle of it one time.

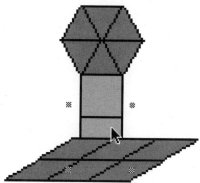

   The small gray "selection" squares show that the bottom orange square is "selected." You can choose menu items to apply to selected shapes.

   b. Choose **Bring To Front** from the **Edit** menu.

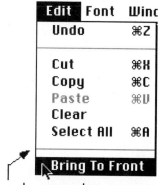

Point to the menu you want and press the mouse button . . .

. . . then move the cursor to **Bring To Front.**

   The orange square is brought to the front of the picture.

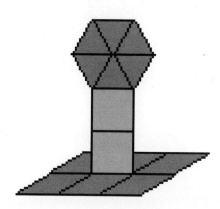

Appendix: Shapes Teacher Tutorial ■ **133**

☞ 3. Glue the tree and ground cover together.

In Step 2, you selected a single shape, the orange square, and applied the action (**Bring To Front**) to it. You can also select a *group* of shapes and apply a tool or action to the entire group at one time. This will make gluing all these shapes together easier.

a. Place the arrow at the top left corner of the group of shapes in the tree.

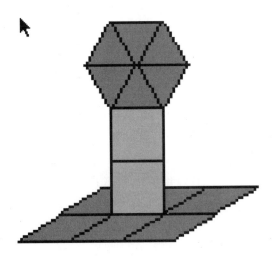

b. Drag diagonally to enclose the shapes in a dotted rectangle . . .

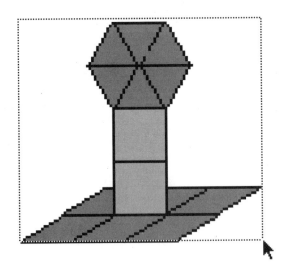

134 ■ *Appendix:* Shapes *Teacher Tutorial*

c. ... and release the mouse button.

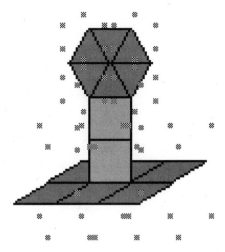

d. Use the Glue tool to glue all the shapes into a group at once by clicking in the middle of any one of them.

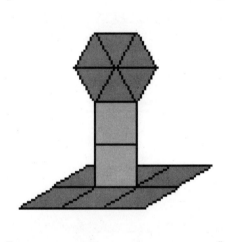

Click on the Arrow tool in the Tool bar. Now the small gray "selection" squares will surround a whole group instead of each individual shape.

You can now do something to all these shapes at once: duplicate them, slide them, turn them, or flip them as one shape. Note that if you click on one of the selected group and slide the whole group, there may be a delay while the *Shapes* program builds an outline of the group.

Next let's try duplicating a group.

☛ 1. Duplicate the tree.

Use the Duplicate tool to make several copies of the tree and place them where you like.

☛ 2. Add any finishing touches.

In the picture below, blue rhombuses connecting the building and the walkway were added, some shapes were moved, and the **Bring To Front** command was applied to add a few finishing touches.

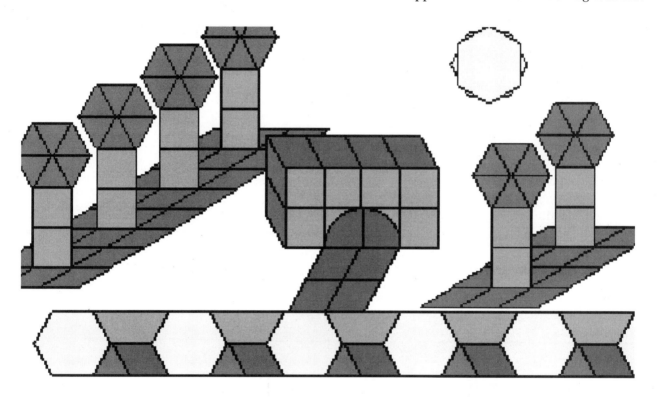

The picture is finished.

136 ■ *Appendix:* Shapes *Teacher Tutorial*

Using *Shapes*

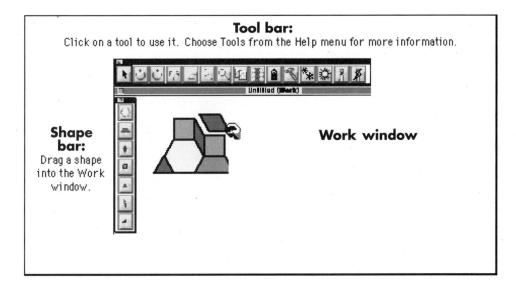

Tool Bar, Shape Bar, and Work Window

Begin by dragging a shape from the Shape bar (the vertical, "floating" bar on the left) to the Work window (the large blank window). Dragging means clicking on a shape and then holding the mouse button down while you move the mouse.

Move the cursor to a shape with the mouse. It becomes a hand.

Click the mouse button and hold it down . . .

. . . while you move the new shape where you want it.

Once the shape is placed in the Work window, you can slide it again by dragging it with the Arrow tool. If you place one shape so that one of its sides is close to a side of another shape, the two shapes will "snap" together.

You can change the position of the shape, or duplicate it, by using the tools in the Tool bar. The tool that is "active," or in use, is surrounded by a black outline (like the arrow tool shown in the diagram on p. 138). Another way to see which tool is active is by the shape of the cursor.

Appendix: Shapes Teacher Tutorial ■ **137**

Only the most commonly used tools are available and displayed for each activity. All tools are available for Free Explore.

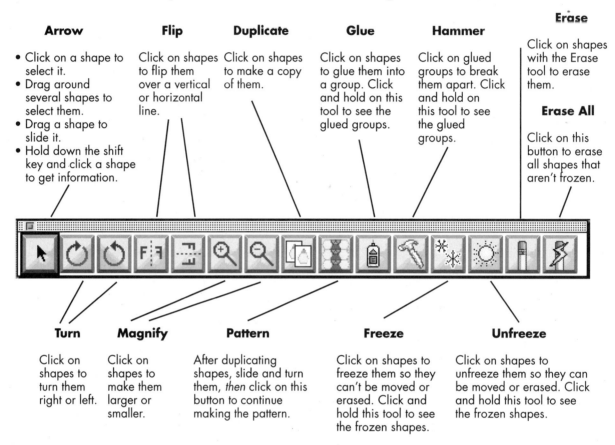

Arrow
- Click on a shape to select it.
- Drag around several shapes to select them.
- Drag a shape to slide it.
- Hold down the shift key and click a shape to get information.

Flip
Click on shapes to flip them over a vertical or horizontal line.

Duplicate
Click on shapes to make a copy of them.

Glue
Click on shapes to glue them into a group. Click and hold on this tool to see the glued groups.

Hammer
Click on glued groups to break them apart. Click and hold on this tool to see the glued groups.

Erase
Click on shapes with the Erase tool to erase them.

Erase All
Click on this button to erase all shapes that aren't frozen.

Turn
Click on shapes to turn them right or left.

Magnify
Click on shapes to make them larger or smaller.

Pattern
After duplicating shapes, slide and turn them, *then* click on this button to continue making the pattern.

Freeze
Click on shapes to freeze them so they can't be moved or erased. Click and hold this tool to see the frozen shapes.

Unfreeze
Click on shapes to unfreeze them so they can be moved or erased. Click and hold this tool to see the frozen shapes.

To use most tools (except Pattern and Erase All, which are buttons):

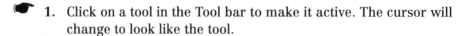 1. Click on a tool in the Tool bar to make it active. The cursor will change to look like the tool.

2. Click in the middle of a shape to perform the action. If you click one of several shapes that are "selected," the action is performed on each of the selected shapes. See the following section, The Arrow Tool, for more information about selecting shapes.

Pattern and Erase All are **buttons**. Simply clicking on a button will perform the action immediately.

The following sections discuss the tools in more detail.

Arrow Tool: Dragging and Selecting

With the Arrow tool, you can drag shapes to slide them. This is the most important use of the Arrow tool.

1. Click in the middle of a shape and hold the mouse button down . . .

2. . . . while you move the mouse, sliding the shape.

3. Release the button to stop sliding.

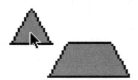

You can also select shapes with the Arrow tool.

Note: You can do most tasks in *Shapes*, including sliding, without ever selecting shapes. It's usually just a convenience for taking some action on several shapes at once. Before we discuss how to select shapes, let's describe what "selecting" means.

Selected shapes are shown surrounded with small gray squares.

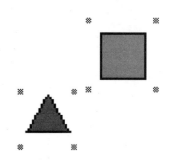

Selected shapes can be copied to the clipboard (a place in computer memory for temporary, invisible storage) or cut—copied to the clipboard *and* removed from the Work window—using commands on the **Edit** menu (see p. 145). Also, if you apply a tool to one shape that is part of a group of selected shapes, the tool will automatically be applied to each shape in the group.

There are two ways to select shapes. First, you can click on any shape with the Arrow tool. That selects the shape. (If it is *already* selected, this will "unselect" the shape.) Second, you can select multiple shapes that are near one another:

1. Place the arrow at one corner of the group of shapes. Press the mouse button.

2. Drag diagonally to enclose the shapes in a dotted rectangle . . .

3. . . . and release the mouse button.

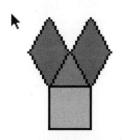  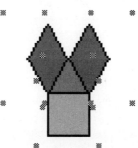

Appendix: Shapes Teacher Tutorial

Now you can do something to all these shapes at once; for example, copy or cut them, slide them all together, or flip them all. Note that if you click on one of the selected group and slide the whole group, there may be a delay while the *Shapes* program builds an outline. Hold the mouse button down without moving the mouse until the outline appears.

You can use the Arrow tool to shift-click on a shape to get information about it. To shift-click, hold the shift key down while clicking on a shape. Click again to clear the message.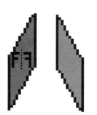

One final feature: If you're using any other tool and you want to use the Arrow tool for a quick selection or slide, just hold down the Command key. *Shapes* will know to use the Arrow tool while the Command key is held down. When you let go of the Command key, *Shapes* will return to the previous tool.

Turn, Flip, and Magnifying Tools: Moving and Sizing

You can use these tools to turn or flip shapes:

One shape: Click on a shape with the tool. For example, if you click on the shape with the first flip tool, the shape flips over a vertical line through the center of the shape.

Several shapes: After the shapes have been selected, click on one of them with the tool. For example, if you click on one with the first Turn tool, each shape turns right around its center.

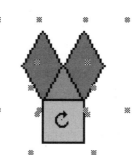
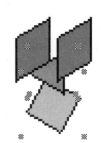

To magnify shapes: If you click on a shape with the first Magnify tool, the shape gets bigger. The second Magnify tool will make it smaller. Shapes that are different sizes will not snap to each other.

140 ■ Appendix: Shapes *Teacher Tutorial*

You can use the Glue tool to glue several shapes together into a group. This group is a new shape you have created. You can slide, turn, and flip it as a unit—that is, as if it were a single shape. For example, you can glue several shapes and then move them or duplicate them.

Glue and Hammer Tools: Combining and Breaking Apart

To use the Glue tool,

 1. Click on Glue tool in the Tool bar to make it active.

 2. Click on each shape you wish to glue together into a group.

   If there are only two shapes, or if two or more shapes are "snapped" or touching, you still have to click on each of them. Click in the middle of each shape and the computer glues them together. Similarly, if you select a group of shapes, and click on one shape, the group will be glued.

   You can add more shapes to an already glued group. Click on one shape in the glued group, then click on one shape in the new group. The two groups will now be glued.

 3. Click on the Arrow tool or any other tool. All the shapes you glued will become a single, new group.

The small gray "selection" squares will surround a whole group instead of each individual shape.

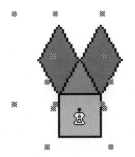 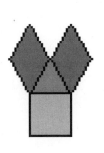

Appendix: Shapes Teacher Tutorial ■ **141**

The group will now act as a single unit. For example, if you click on the group with the Right Turn tool, the group turns *as one shape* around the center of the group.

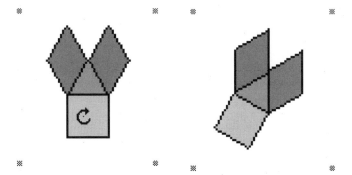

What if you want to make two separate groups? Suppose you made two different kinds of houses and you want each to be glued in a different group. You can't glue all the shapes in each of the houses at once; that would make one two-house group. Instead, you must glue one house, clicking on the Arrow tool to end the gluing process and glue that group. Then you must choose the Glue tool again and glue the second house.

Sometimes it helps to know what shapes are already glued into groups. Hold the mouse button down on either the Glue or Hammer tool on the Tool bar to see which shapes are in which groups.

 1. Hold the button down on the Glue (or Hammer) tool to see the group number on each shape.

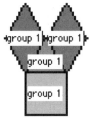

Use the Hammer tool to break apart glued shapes. Click on any shape in the group with the hammer to break apart the group.

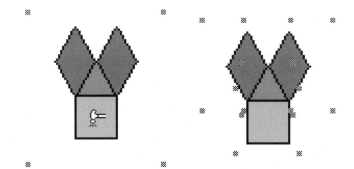

142 ■ *Appendix: Shapes Teacher Tutorial*

The Pattern tool is a **button**. You just click on it to perform the action (the Erase All button works the same way). To use the Pattern button, you must first make a pattern.

☞ 1. Make a basic unit for the pattern. Any shape or combination of shapes can be used as this unit. If you plan to turn this unit as part of your pattern, you must glue the shapes in the basic unit to form a group. Turn patterns must have a single glued group as the basic unit.

☞ 2. Duplicate this unit with the Duplicate tool. (You can also choose **Copy** and then **Paste** from the **Edit** menu.)

☞ 3. Move the duplicate of the unit where you want it to be for the start of the pattern by using the tools. You can slide the duplicate or turn it. (Remember, if you turn it as in this example, all the shapes in the duplicate must have been glued into one group.)

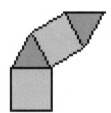

☞ 4. Now, click on the Pattern button as many times as you wish to continue the pattern. If a duplicate goes off the Work window, you will hear a beep and the Pattern button will not make any more copies.

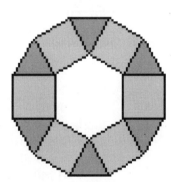

Duplicate Tool and Pattern Button: Making Repeating Patterns

To make the pattern below, glue a hexagon, triangle, and blue rhombus to form the basic unit. Duplicate this unit; slide it to the right; then click on the Pattern button.

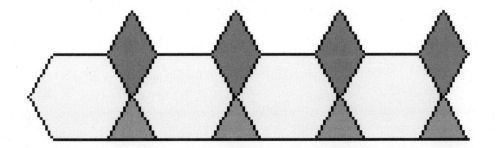

Appendix: Shapes *Teacher Tutorial* ▪ **143**

The Pattern tool makes patterns composed of slides and turns. If you want flipped shapes in your pattern, use these steps to make a basic unit (Step 1, p. 143):

   a. Glue a group.
   b. Duplicate the group.
   c. Use the Flip tool to flip the duplicate.
   d. Glue the two groups (the original and its flipped [mirror] image) together to make the basic unit for the pattern.
   e. Continue with Step 2 (p. 143).

Freeze and Unfreeze Tools: Combining and Breaking Apart

You can use the Freeze tool to "freeze" shapes. Frozen shapes can't be moved or erased. They are frozen in place (in comparison, the Glue tool glues shapes to each other; however, the glued group can still be moved).

Freezing shapes allows you to manipulate other shapes without accidentally moving or erasing those that are frozen. Click on a shape with the Unfreeze tool to unfreeze it. Click and hold either tool on the Tool bar to see the frozen shapes.

Using Menu Commands

To use any menu commands, do the following:

Point to the menu you want and press the mouse button . . .

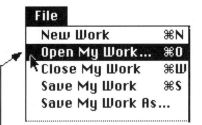

. . . then drag the selection bar to your choice and release the button.

Shapes Menus

The **File** menu deals with documents and quitting.

About Menus

New Work starts a new document. The ⌘N indicates that, instead of selecting this from the menu, you could enter ⌘N or **Command N** by holding down the **Command** key (with the ⌘ and symbols on it) and then pressing the N key.

Open My Work opens previously saved work.

Close My Work closes present work.

Save My Work saves the work.

Save My Work As saves the work with a new name or to a different disk or folder.

Change Activity lets you choose a different activity.

Page Setup allows you to set up how the printer will print your work.

Print prints your work.

Quit quits *Shapes*.

File	
New Work	⌘N
Open My Work...	⌘O
Close My Work	⌘W
Save My Work	⌘S
Save My Work As...	
Change Activity...	
Page Setup...	
Print...	⌘P
Quit	⌘Q

When you save your work for the first time, a dialogue box opens. Type a name. You may wish to include your name or initials, your work, and the date. For the remainder of that session, you can save your work simply by selecting the **Save** menu item or pressing ⌘S (**Command S**).

To share the computer with others, save your work then choose **Close My Work**. Later, to resume your work, choose **Open My Work** and select the work you saved.

When you **Quit**, you are asked whether you wish to save your work. If you choose to save at that time (you don't have to if you just saved), the same steps are followed.

The **Edit** menu contains choices to use when editing your work.

Cut deletes the selected object and saves it to a space called the clipboard.

Copy copies selected object on the clipboard.

Paste puts the contents of the clipboard into the Work window.

Clear deletes the selected object but does not put it on the clipboard.

Select All selects all shapes. This is not only a fast and handy shortcut, it also helps when some shapes are nearly or totally off the Work window.

Bring To Front puts the selected shapes "in front of" unselected shapes. That is, each new shape you create will be "in front of" the shapes already on the Work window. If you want to change this, select a shape that is in the "back," hidden by other shapes, and choose **Bring To Front**.

Edit	
Undo	⌘Z
Cut	⌘X
Copy	⌘C
Paste	⌘V
Clear	
Select All	⌘A
Bring To Front	

Appendix: Shapes Teacher Tutorial

Font
Chicago
Courier
Geneva
Helvetica
Monaco
New York
Times
Size ▶
Style ▶
All Large

The **Font** menu is used to change the appearance of text in the Show Notes window. In order for any command in the **Font** menu to be highlighted, the Show Notes command in the **Windows** menu must be open.

The first names are choices of typeface. They will vary according to the fonts available in your computer.

Size and **Style** have additional choices; pull down to select them and then to the right. See the example for **Style** shown at left. The **Size** choice works the same way.

All Large changes all text in all windows to a large-size font. This is useful for demonstrations. This selection toggles (changes back and forth) between **All Large** and **All Small**.

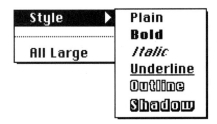

Windows
Hide Tools
Show Shapes
Show Work
Hide Notes

Shapes
All Large
All Medium
All Small

The **Windows** menu shows or hides the windows. If you hide a window such as the Tool bar, the menu item changes to **Show** followed by the name of the window—for example, **Show Tools**. You can also hide a window by clicking in the "close box" in the upper-left corner of the window.

The **Notes** Window is designed to be a word processor. Students might use this to record a strategy they used to solve a problem, or to write notes to themselves as reminders of how they plan to continue in the next math session.

The **Shapes** menu contains commands to change the size of the shapes.

All Large, **All Medium**, and **All Small** changes the size of all the shapes. To change the size of just some of the shapes, use the Magnify tools. Note that shapes of different sizes do not snap to each other.

146 ■ *Appendix:* Shapes *Teacher Tutorial*

The **Number** menu is available for only certain activities that have more than one task. Select a number off the submenu to work on that number task.

The **Options** menu allows you to customize *Shapes*.

Vertical Mirror and **Horizontal Mirror** toggle (turn on and off) the Work windows mirrors, which reflect any action you take.

Snap toggles the "snap" feature, in which shapes, when moved, automatically "snap" or move next to, any other shape they are close to. You may want to turn that feature off if you have lots of shapes on the Work window, or if you are copying many shapes. This feature can slow down moving the shapes. You can turn it off temporarily to speed things up.

The **Help** menu provides assistance.

Windows provides information on the three main windows: the Tool bar, the Shape bar, and the Work window.

Tools provides information on tools (represented on the Tool bar as icons, or pictures).

Directions provides instructions for the present activity.

Hints gives a series of hints on the present activity, one at a time. It is dimmed when there are no available hints.

Appendix: Shapes Teacher Tutorial ■ **147**

Trouble-Shooting

This section contains suggestions for how to correct errors and what to do in some troubling situations.

If you are new to using the computer, you might also ask a computer coordinator or an experienced friend for help.

No *Shapes* Icon to Open

- Check that *Shapes* has been installed on your computer by looking at a listing of the hard disk.
- Open the folder labeled *Shapes* by double-clicking on it.
- Find the icon for the *Shapes* application and double-click on it.

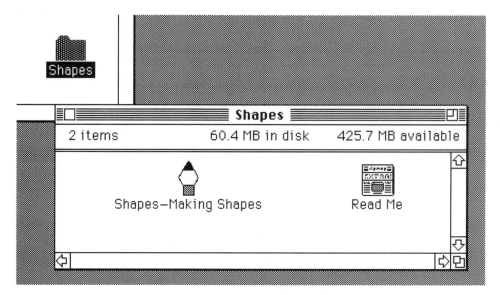

Nothing Happened After Double-Clicking on the *Shapes* Icon

- If you are sure you double-clicked correctly, wait a bit longer. *Shapes* takes a while to open or load and nothing new will appear on the screen for a few seconds.
- On the other hand, you may have double-clicked too slowly, or moved the mouse between your clicks. In that case, try again.

In the Wrong Activity

- Choose **Change Activity** from the **File** menu.

A Window Closed by Mistake

- Choose **Show Window** from the **Windows** menu.

Windows or Tools Dragged to a Different Position by Mistake

- Drag the window back into place by following these steps: Place the pointer arrow in the stripes of the title bar. Press and hold the button as you move the mouse. An outline of the window indicates the new location. Release the button and the window moves to that location.

I Clicked Somewhere and Now *Shapes* Is Gone! What Happened?

You probably clicked in a part of the screen not used by *Shapes* and the computer therefore took you to another application, such as the "desktop."

- Click on a *Shapes* window, if visible.
- Double-click on the *Shapes* program icon.

How Do I Select a Section of Text?

In certain situations, you may wish to copy or delete a section or block of text in the Notes window.

- Point and click at one end of the text. Drag the mouse by holding down the mouse button as you move to the other end of the text. Release the mouse button. Then use the **Edit** menu to **Copy**, **Cut**, and **Paste**.

System Error Message

Some difficulty with the *Shapes* program or your computer caused the computer to stop functioning.

- Turn off the computer and repeat the steps to turn it on and start *Shapes* again. Any work that you saved will still be available to open from your disk.

I Tried to Print and Nothing Happened

- Check that the printer is connected and turned on.
- When printers are not functioning properly, a system error may occur, causing the computer to "freeze." If there is no response from the keyboard or when moving or clicking with the mouse, you may have to turn off the computer and start over.

I Printed the Work Window but Not Everything Printed

- Choose the Color/Grayscale option for printing.
- If your printer has no such option (e.g., an older black-and-white printer), you need to find a different printer to print graphics in color.

Appendix: Shapes *Teacher Tutorial* ▪ **149**

Shapes Messages

If the *Shapes* program does not understand a command or has a suggestion, a dialogue box may appear with one of the following messages. Read the message, click on **[OK]** or press **<return>** from the keyboard, and correct the situation as needed.

Disk or directory full.

The computer disk is full.

- Use **Save My Work As** to choose a different disk.

I'm having trouble with the disk or drive.

The disk might be write-protected, there is no disk in the drive, or some similar problem.

- Use **Save My Work As** to choose a different disk.

Out of space.

There is no free memory left in the computer.

- Eliminate shapes you don't need.
- Save and start new work.

How to Install *Shapes* on Your Computer

The *Shapes* disk that you received with this unit contains the *Shapes* program and a Read Me file. You may run the program directly from this disk, but it is better to put a copy of the program and the Read Me file on your hard disk and store the original disk for safekeeping. Putting a program on your hard disk is called *installing* it.

Note: *Shapes* runs on a Macintosh II computer or above, with 4 MB of internal memory (RAM) and Apple System Software 7.0 or later. (*Shapes* can run on a Macintosh with less internal memory, but the system software must be configured to use a minimum of memory.)

To install the contents of the *Shapes* disk on your hard drive, follow the instructions for your type of computer or these steps:

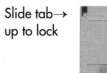

Slide tab→ up to lock

Back of disk

☞ 1. Lock the *Shapes* program disk by sliding up the black tab on the back, so the hole is open.

The *Shapes* disk is your master copy. Locking the disk allows copying while protecting its contents.

☞ 2. Insert the *Shapes* disk into the floppy disk drive.

☞ 3. Double-click on the icon of the *Shapes* disk to open it.

☞ 4. Double-click on the Read Me file to open and read it for any recent changes in how to install or use *Shapes*. Click in the Close box after reading.

☞ 5. Click on and drag the *Shapes* disk icon (the outline moves) to the hard disk icon until the hard disk icon is highlighted, then release the mouse button.

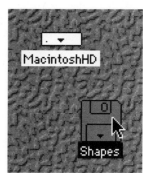

The message appears indicating that the contents of the *Shapes* disk are being copied to the hard disk. The copy is in a folder on the hard disk with the name *Shapes*.

☞ 6. Eject the *Shapes* disk by selecting it (clicking on the icon) and choosing **Put Away** from the **File** menu. Or, drag the icon into the trash. Store the disk in a safe place.

☞ 7. If the hard disk window is not open on the desktop, open the hard disk by double-clicking on the icon.

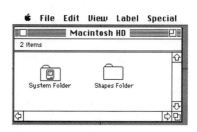

When you open the hard disk, the hard disk window appears, showing you the contents of your hard disk. It might look something like this. Among its contents is the folder labeled *Shapes* holding the contents of the *Shapes* disk.

Appendix: Shapes Teacher Tutorial ▪ **151**

☞ 8. Double-click the *Shapes* folder to select and open it.

When you open the *Shapes* folder, the window contains the program and the Read Me file.

To select and run *Shapes,* double-click on the program icon.

Optional

For ease at startup, you might create an alias for the *Shapes* program by following these steps:

☞ 1. Select the program icon.

☞ 2. Choose **Make Alias** from the **File** menu.

The alias is connected to the original file that it represents, so that when you open an alias, you are actually opening the original file. This alias can be moved to any location on the desktop.

☞ 3. Move the *Shapes* alias out of the window to the desktop space under the hard disk icon.

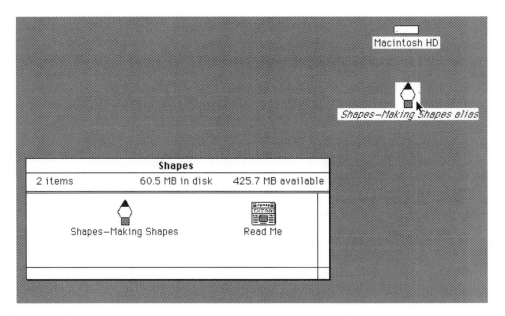

For startup, double-click on the *Shapes* alias instead of opening the *Shapes* folder to start the program inside.

152 ■ *Appendix:* Shapes *Teacher Tutorial*

Other Disk Functions

For classroom management purposes, you might want to save student work on a disk other than the hard drive. Make sure that the save-to disk has been initialized (see instructions for your computer system).

Saving Work on a Different Disk

☞ 1. Insert the save-to disk into the drive.

☞ 2. Choose **Save My Work As** from the **File** menu.

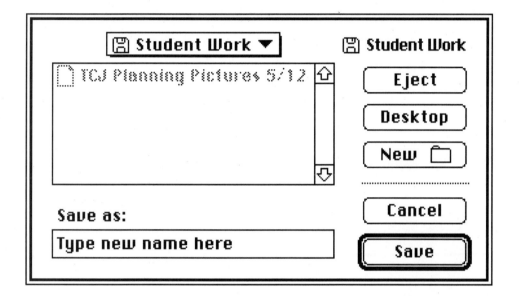

The name of the disk the computer is saving to is displayed in the dialogue box. To choose a different disk, click the **[Desktop]** button and double-click to choose and open a disk from the new menu.

☞ 3. Type a name for your work if you want to have a new or different name from the one it currently has.

☞ 4. Click on **[Save]**.

As students no longer need previously saved work, you may want to delete their work (called "files") from a disk. This cannot be accomplished from inside the *Shapes* program. However, you can delete files from disks at any time by following directions for how to "Delete a File" for your computer system.

Deleting Copies of Student Work

Appendix: Shapes Teacher Tutorial ■ **153**

Blackline Masters

Family Letter   156

Investigation 1
Shape Cutouts A–F   157
Pattern Block Cutouts   160

Investigation 2
Pattern Block Puzzles 1–10   166

Investigation 3
Student Sheet 1, Shape Hunt   176
Shape Hunt at Home   177

Investigation 4
Make-a-Shape Cards   178
Student Sheet 2, Fill the Hexagons Gameboard   181

Investigation 5
Geoblock Match-Up Gameboards 1–6   182

General Resources for the Unit
Choice Board Art   188

_____, 19____

Dear Family,

Our class is beginning a new unit in mathematics called *Making Shapes and Building Blocks.* For the next few weeks, while we are investigating geometry, your child will "make shapes" like triangles and rectangles, and "build a block" just like another one in our Geoblock set. In these and many other ways, we will explore both two- and three-dimensional shapes.

We will look for examples of shapes in the everyday world—for example, a kite is often a *rhombus* or diamond shape, and a book is usually a *rectangle.* We'll put our ideas into a Book of Shapes and a shape mural. The children will also go on a Shape Hunt looking for three-dimensional shapes. They may find a basketball as an example of a *sphere,* and a soup can or rolling pin as an example of a *cylinder.*

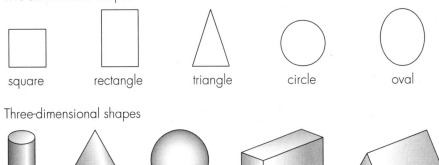

You can help your child by looking for opportunities to talk about shapes.

- Look for different shapes in the environment, at home or while you are out. You can look for both two-dimensional and three-dimensional shapes. Encourage your child to look closely and describe what each shape looks like.

- Making shapes is a good way to learn about them. At home, your child might use clay or playdough, building blocks, drinking straws, or a loop of yarn or rope to make different shapes. Drawing shapes is also fun. Your child might like to design pictures using shapes, as we will be doing in class.

- You and your child might visit the children's section of the local library and find books about shapes to read together.

Have a good time exploring these ideas with your child.

Sincerely,

SHAPE CUTOUTS A AND B

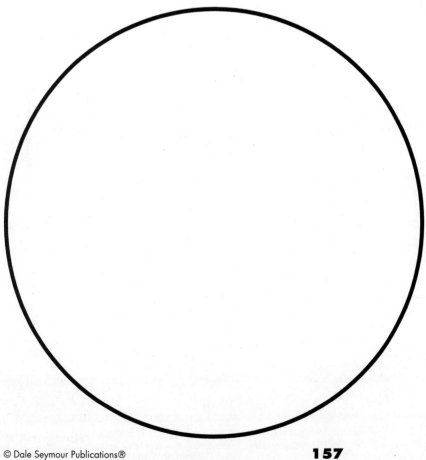

Investigation 1
Making Shapes and Building Blocks

SHAPE CUTOUTS C AND D

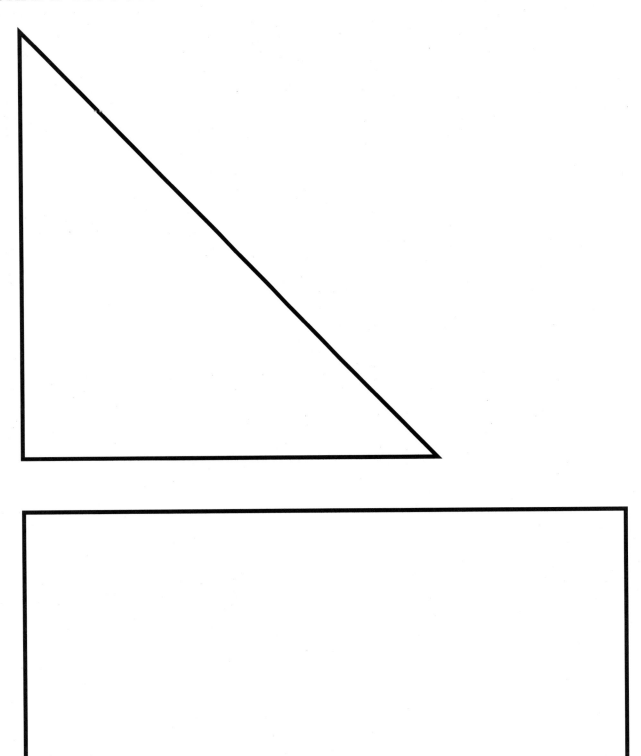

SHAPE CUTOUTS E AND F

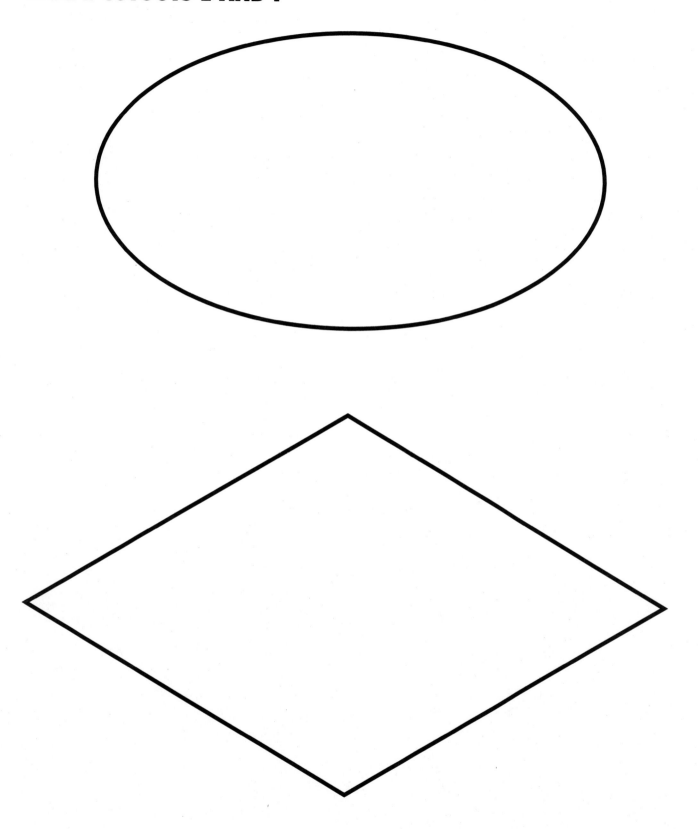

PATTERN BLOCK CUTOUTS (page 1 of 6)

Duplicate these hexagons on yellow paper and cut apart.

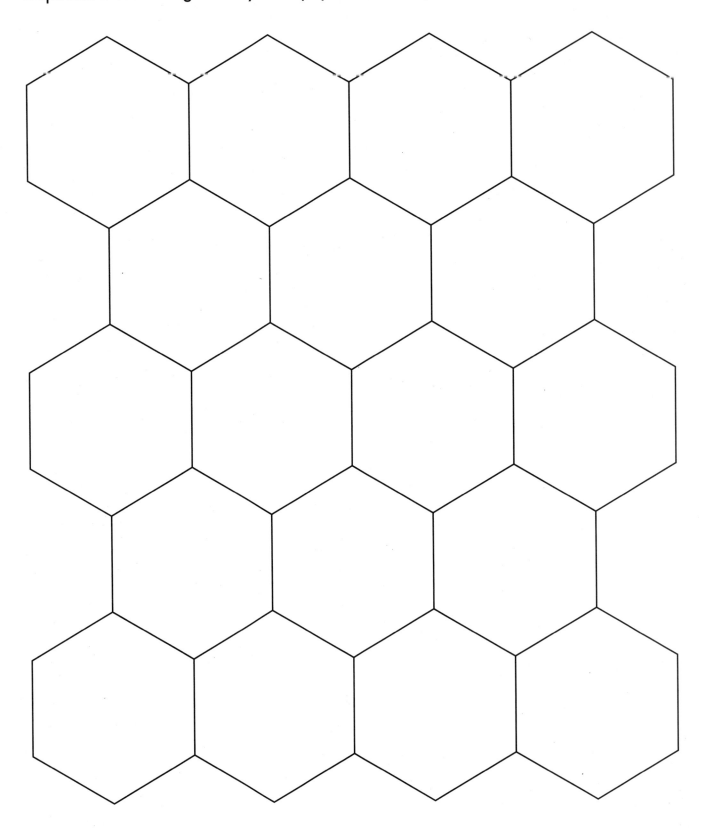

PATTERN BLOCK CUTOUTS (page 2 of 6)
Duplicate these trapezoids on red paper and cut apart.

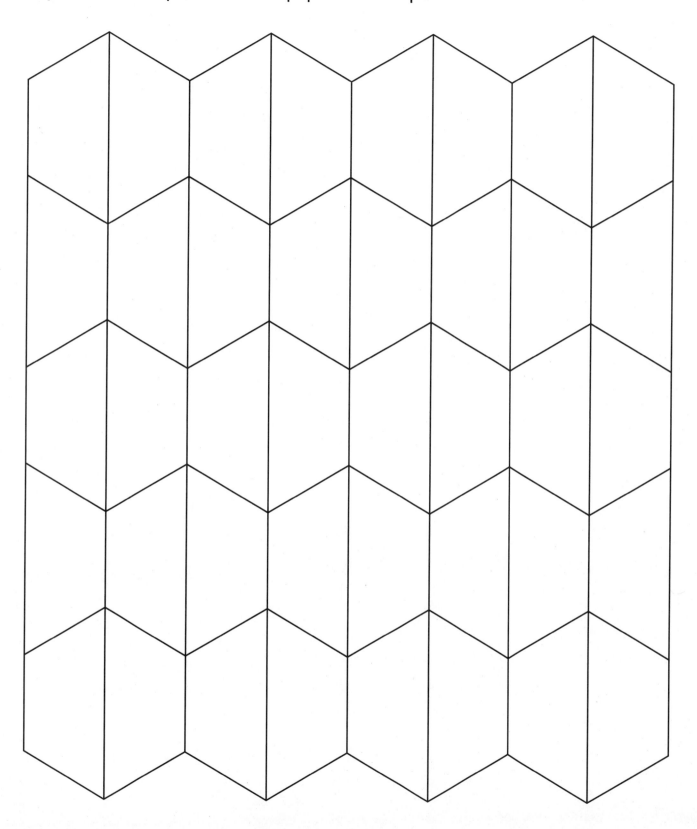

Investigation 1
Making Shapes and Building Blocks

PATTERN BLOCK CUTOUTS (page 3 of 6)

Duplicate these triangles on green paper and cut apart.

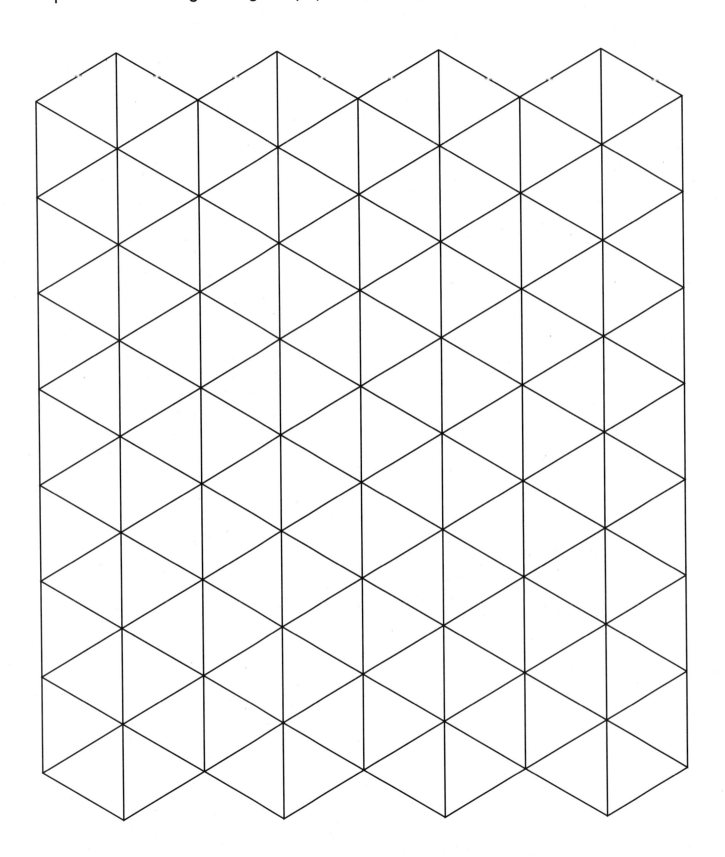

PATTERN BLOCK CUTOUTS (page 4 of 6)

Duplicate these squares on orange paper and cut apart.

PATTERN BLOCK CUTOUTS (page 5 of 6)
Duplicate these rhombuses on blue paper and cut apart.

PATTERN BLOCK CUTOUTS (page 6 of 6)
Duplicate these rhombuses on tan paper and cut apart.

PATTERN BLOCK PUZZLE 1

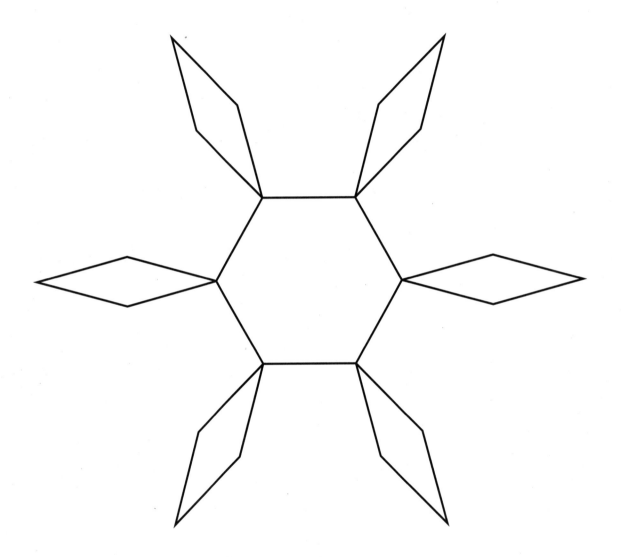

166

*Investigation 2
Making Shapes and Building Blocks*

PATTERN BLOCK PUZZLE 2

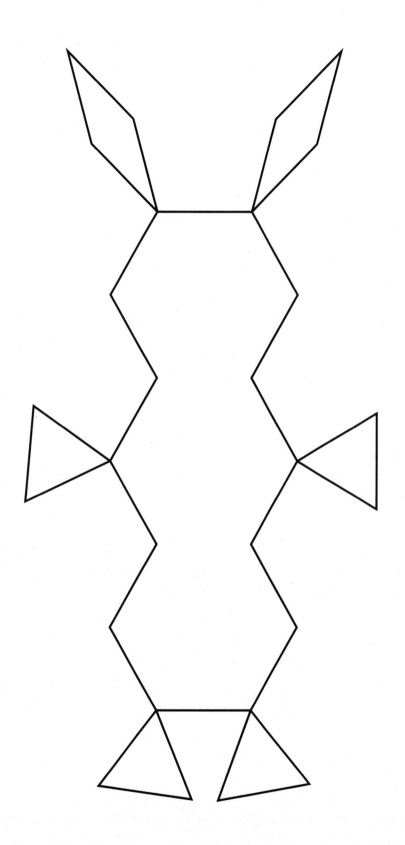

167

Investigation 2
Making Shapes and Building Blocks

PATTERN BLOCK PUZZLE 3

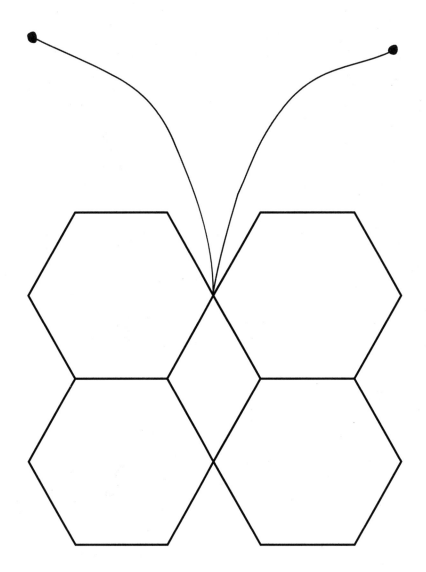

168

Investigation 2
Making Shapes and Building Blocks

PATTERN BLOCK PUZZLE 4

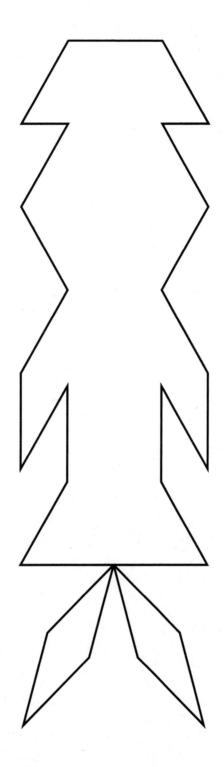

PATTERN BLOCK PUZZLE 5

Investigation 2
Making Shapes and Building Blocks

PATTERN BLOCK PUZZLE 6

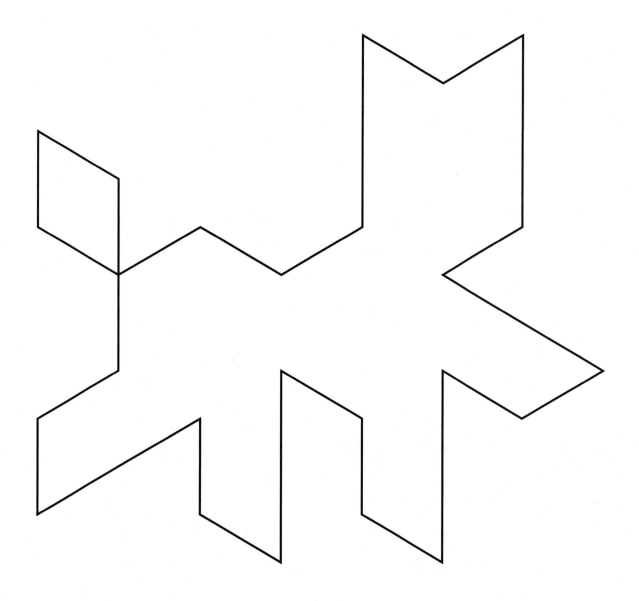

171

Investigation 2
Making Shapes and Building Blocks

PATTERN BLOCK PUZZLE 7

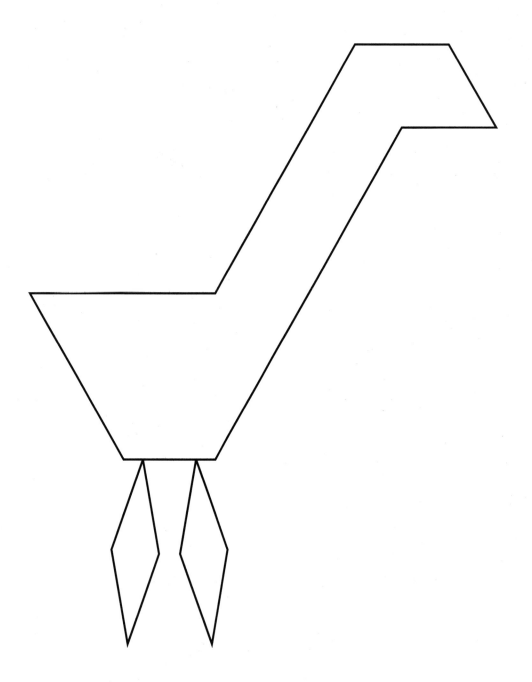

172

Investigation 2
Making Shapes and Building Blocks

PATTERN BLOCK PUZZLE 8

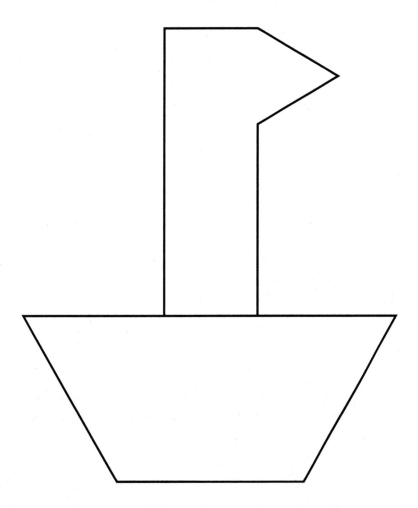

PATTERN BLOCK PUZZLE 9

PATTERN BLOCK PUZZLE 10

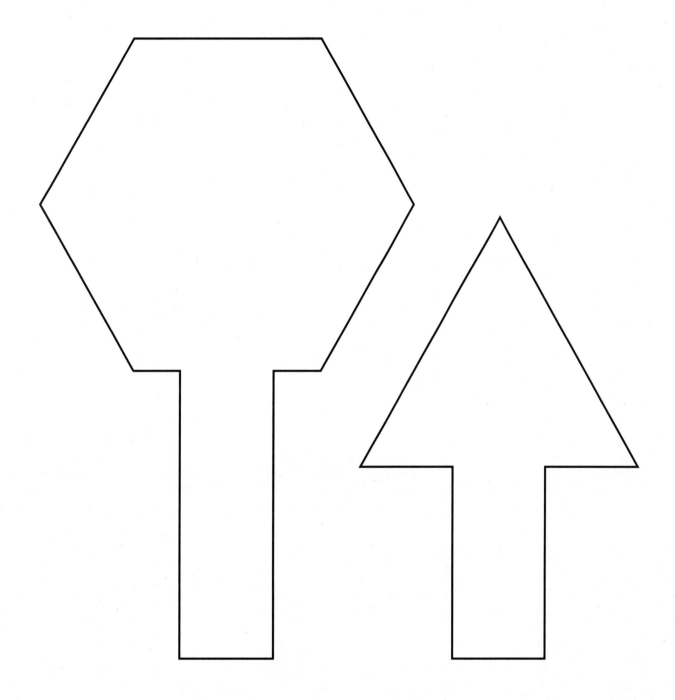

175

Investigation 2
Making Shapes and Building Blocks

Name _____ Date _____

Student Sheet 1

Shape Hunt

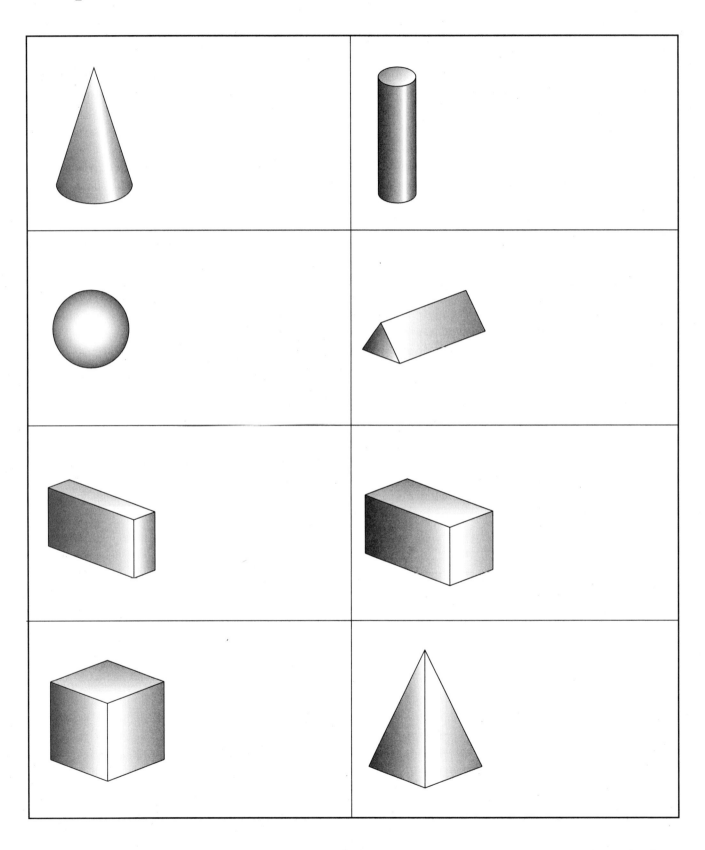

© Dale Seymour Publications®

176

Investigation 3
Making Shapes and Building Blocks

SHAPE HUNT AT HOME

Dear Family,

Your child will be going on a Shape Hunt at home to look for real-world objects that have these three-dimensional shapes. You can help by hunting for these shapes together and writing down all the different examples you find. You might help your child write the words, or write them yourself as your child finds an object.

sphere

cylinder

cube

cone

rectangular prism

MAKE-A-SHAPE CARDS (page 1 of 3)

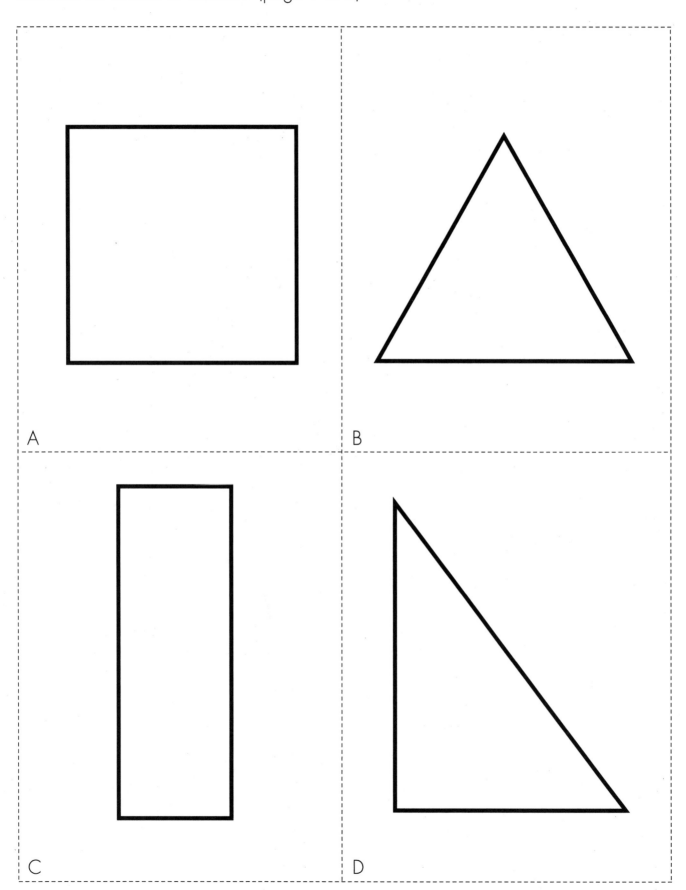

A

B

C

D

Investigation 4
Making Shapes and Building Blocks

MAKE-A-SHAPE CARDS (page 2 of 3)

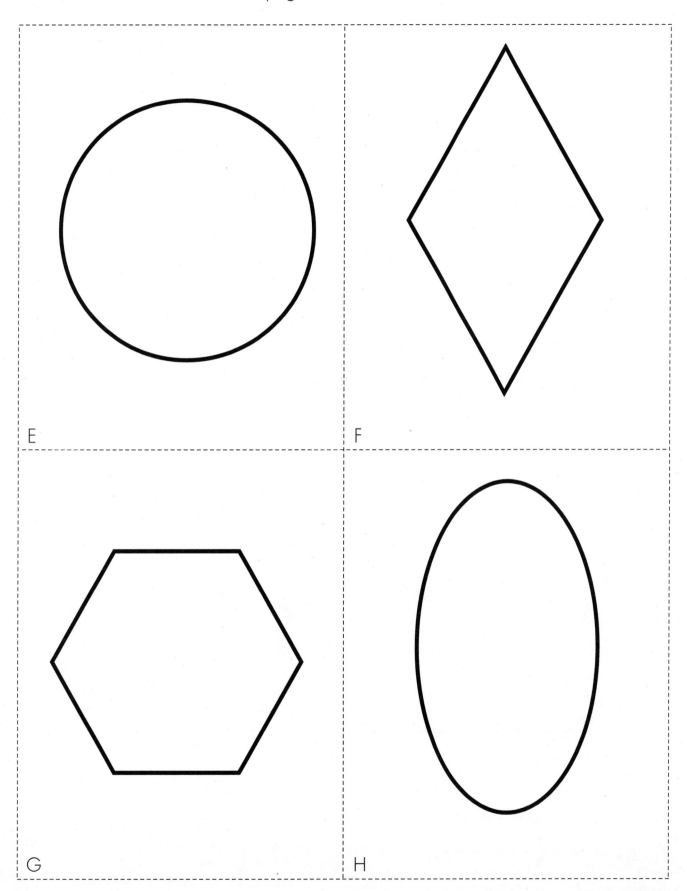

MAKE-A-SHAPE CARDS (page 3 of 3)

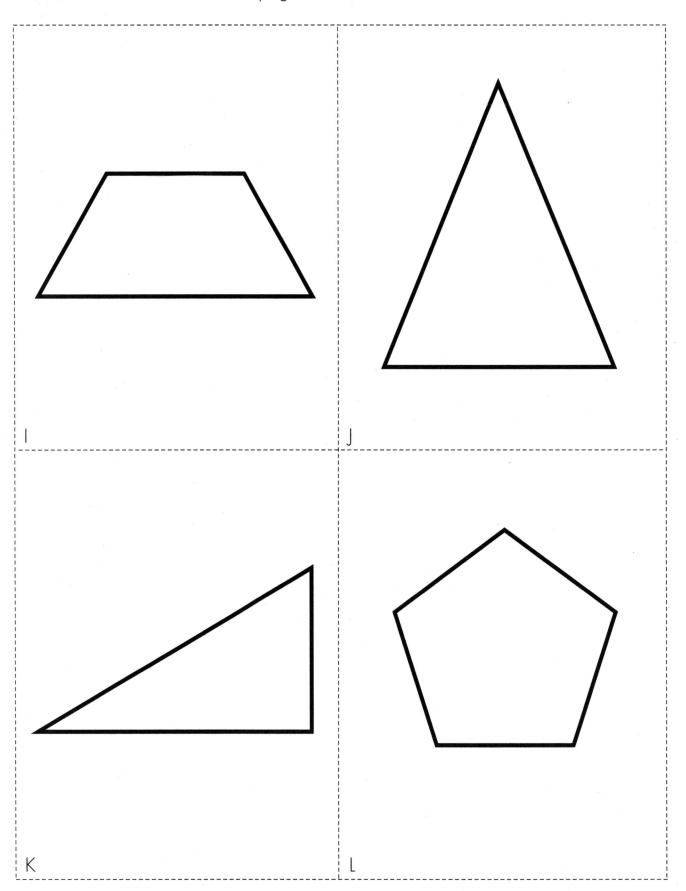

Name _____ Date _____

Student Sheet 2

Fill the Hexagons Gameboard

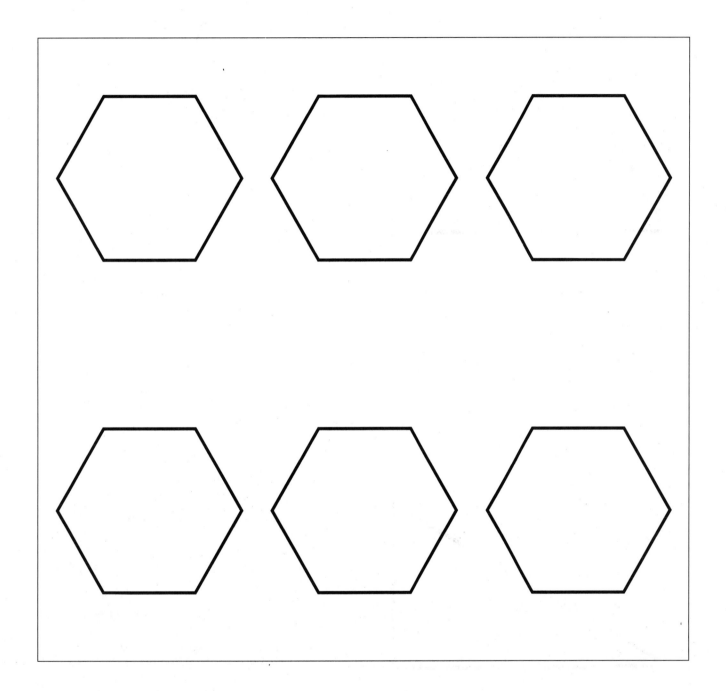

Investigation 4
Making Shapes and Building Blocks

GEOBLOCK MATCH-UP GAMEBOARD 1

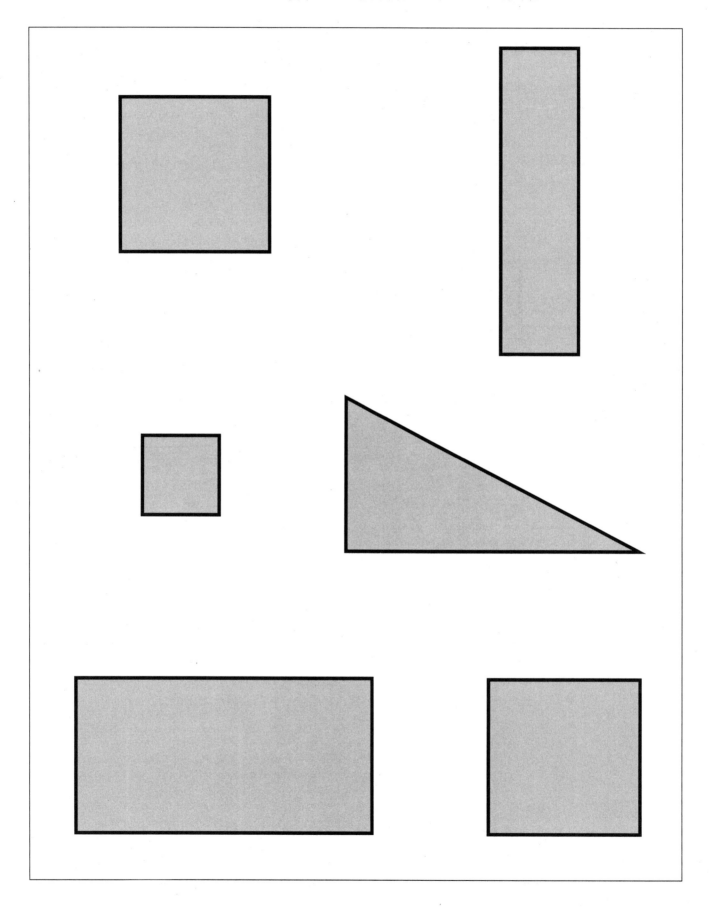

GEOBLOCK MATCH-UP GAMEBOARD 2

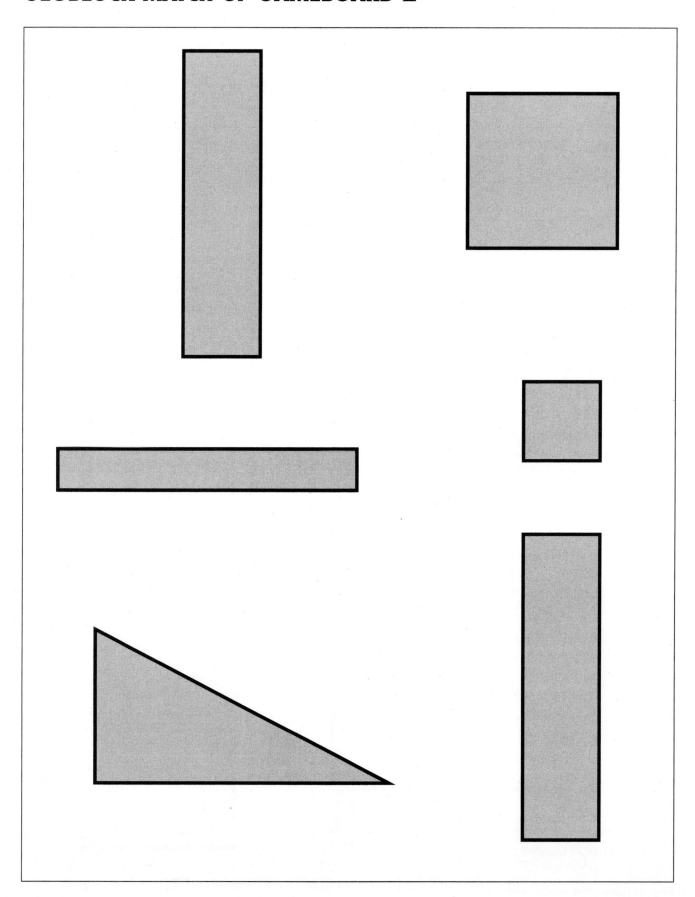

GEOBLOCK MATCH-UP GAMEBOARD 3

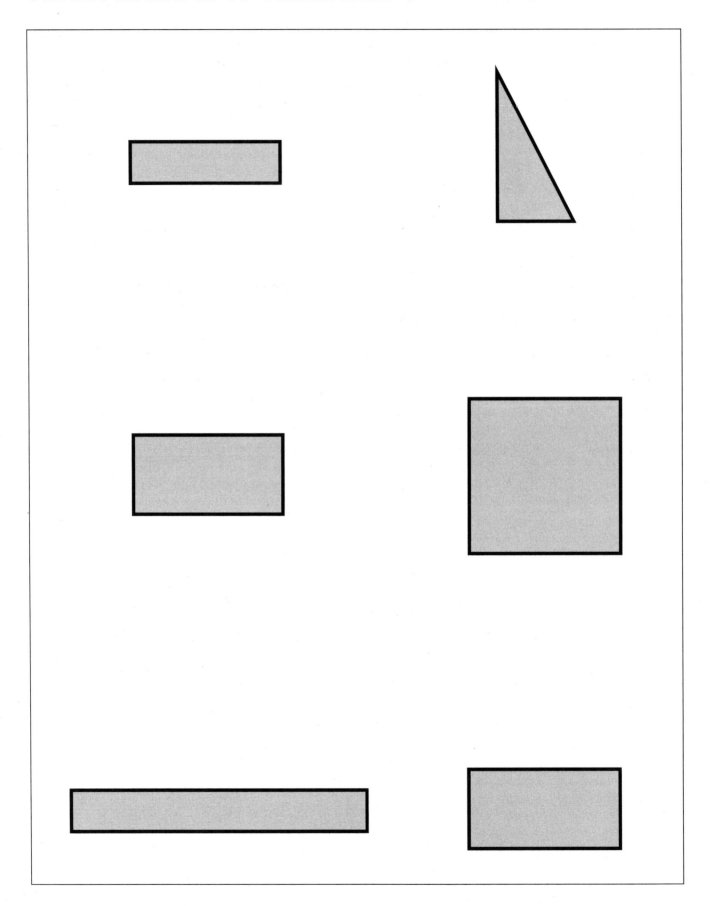

Investigation 5
Making Shapes and Building Blocks

GEOBLOCK MATCH-UP GAMEBOARD 4

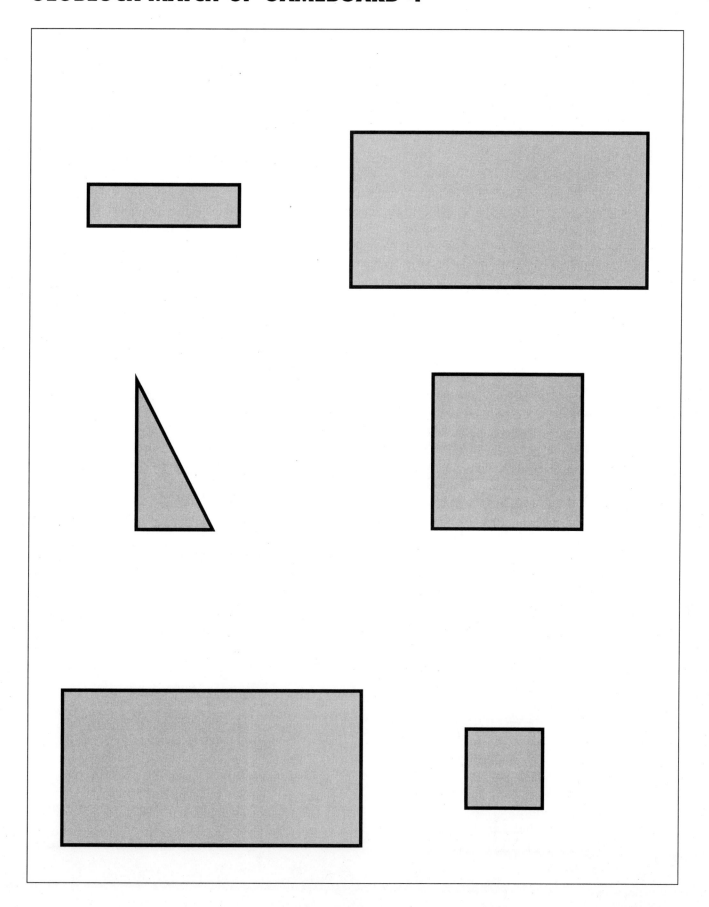

GEOBLOCK MATCH-UP GAMEBOARD 5

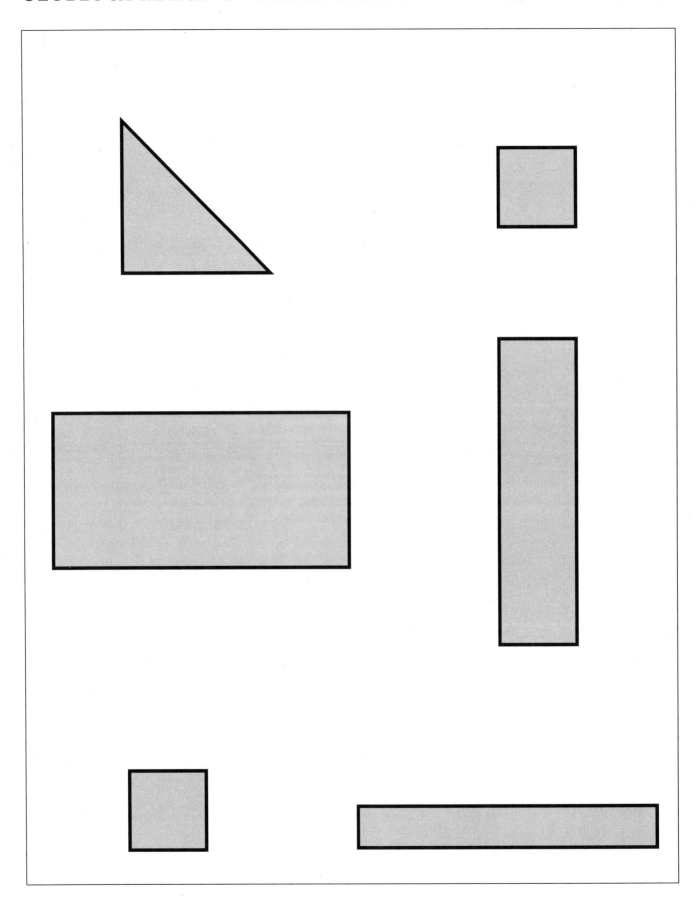

GEOBLOCK MATCH-UP GAMEBOARD 6

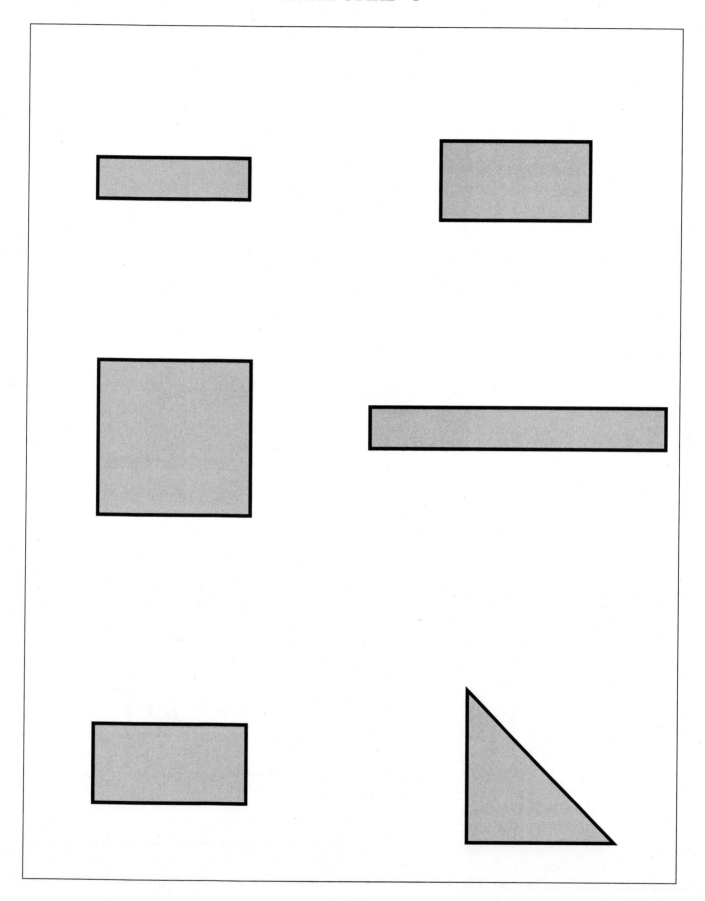

CHOICE BOARD ART (page 1 of 5)

Choice Board art for
Book of Shapes

Choice Board art for
Pattern Block Pictures

Choice Board art for
Shape Mural

CHOICE BOARD ART (page 2 of 5)

Choice Board art for
Free Explore with *Shapes*

Choice Board art for
Pattern Block Puzzles

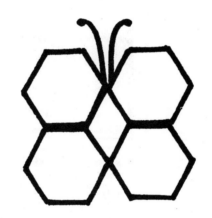

Choice Board art for
Shape Hunt

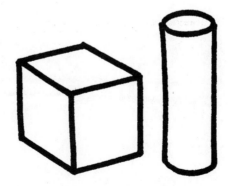

CHOICE BOARD ART (page 3 of 5)

Choice Board art for
Exploring Geoblocks

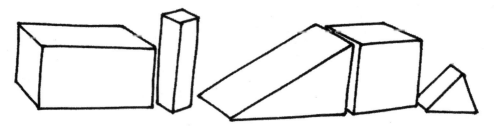

Choice Board art for
The Shape of Things

Choice Board art for
Clay Shapes

CHOICE BOARD ART (page 4 of 5)

Choice Board art for
Fill the Hexagons

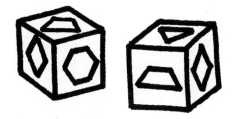

Choice Board art for
Build a Block

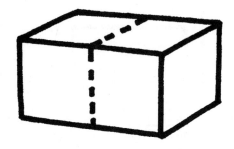

Choice Board art for
Quick Images

CHOICE BOARD ART (page 5 of 5)

Choice Board art for
Matching Faces

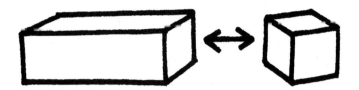

Choice Board art for
Geoblock Match-up

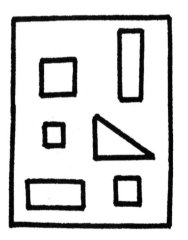

Choice Board art for
Planning Pictures